Stephan Knapek

Specific components of aversive odor memory in Drosophila melanogaster

Stephan Knapek

Specific components of aversive odor memory in Drosophila melanogaster

Synapsin and Bruchpilot, two presynaptic proteins underlying specific phases of olfactory aversive memory in the fruit fly

Südwestdeutscher Verlag für Hochschulschriften

Imprint
Any brand names and product names mentioned in this book are subject to trademark, brand or patent protection and are trademarks or registered trademarks of their respective holders. The use of brand names, product names, common names, trade names, product descriptions etc. even without a particular marking in this work is in no way to be construed to mean that such names may be regarded as unrestricted in respect of trademark and brand protection legislation and could thus be used by anyone.

Publisher:
Südwestdeutscher Verlag für Hochschulschriften
is a trademark of
Dodo Books Indian Ocean Ltd., member of the OmniScriptum S.R.L Publishing group
str. A.Russo 15, of. 61, Chisinau-2068, Republic of Moldova Europe
Printed at: see last page
ISBN: 978-3-8381-2061-4

Zugl. / Approved by: Würzburg, Bayerische Julius-Maximilians Universität, Diss., 2010

Copyright © Stephan Knapek
Copyright © 2011 Dodo Books Indian Ocean Ltd., member of the OmniScriptum S.R.L Publishing group

Contents

1 Introduction 3
    1.1 Classical associative learning and memory 3
    1.2 The fruit fly *Drosophila* as a model organism 4
    1.3 Associative olfactory learning and memory in *Drosophila* 8
    1.4 The CS pathway: olfactory system of *Drosophila melanogaster* 10
    1.5 The US pathways: The role of dopamine and octopamine for mediating punishment or reward 12
    1.6 The mushroom body 13
    1.7 Different components of aversive olfactory memory 17
    1.8 Molecular mechanisms of olfactory learning 19
    1.9 Regulation of presynaptic neurotransmitter release 21
    1.10 Aim of the work 24

2 Material and Methods 25
    2.1 Fly care and genotypes 25
    2.2 Immunohistochemistry 28
    2.3 Olfactory conditioning 29
    2.4 Anesthesia-resistant memory (ARM) 29
    2.5 Responsiveness to electric shocks and odors 30
    2.6 Employed odors 31
    2.7 Statistics 31

3 Results 32
    3.1 Reinforcer intensity differentially affects distinct memory components 32
    3.2 Synapsin is required for short lasting memory 34
    3.3 Synapsin is specifically required for anesthesia-sensitive memory 39
    3.4 Synapsin as a potential target of PKA 40
    3.5 Bruchpilot is preferentially required for ARM 42
    3.6 Bruchpilot requirement for ARM is specific to the mushroom body 45

	3.7	Bruchpilot knock-down reduces immediate memory	49
	3.8	Bruchpilot knock-down and *rutabaga* mutant show additive memory impairment	49

4		**Discussion**	**52**
	4.1	Reinforcer intensity and memory retention	52
	4.2	Synapsin mediates the labile memory ASM	52
	4.3	Synapsin as a potential phosphorylation target of PKA	54
	4.4	Bruchpilot, a new protein for ARM	55
	4.5	Dissociation of ARM and ASM	56

5	**References**	**59**

6	**List of abbreviations**	**75**

7	**Summary**	**77**

8	**Zusammenfassung**	**79**

1 Introduction

1.1 Classical associative learning and memory

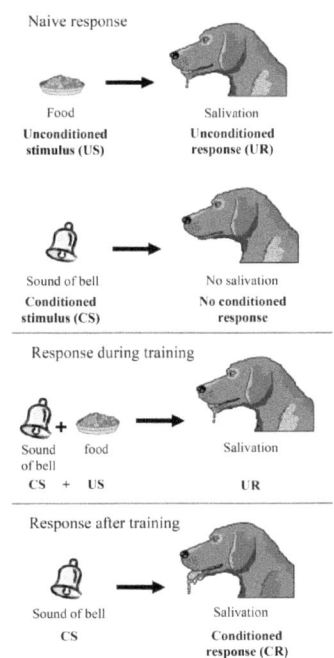

Figure 1: Classical associative learning. A meaningful stimulus (US, e.g. reward or pain) is repeatedly paired with a neutral stimulus (CS, e.g. sound or odor). After such training, the CS is associated with the US and therefore elicits now a conditioned response (CR) similar to the unconditioned response (UR) previously only triggered by the US (figure modified from www.skewsme.com).

Learning is an experience-dependent, enduring change in behavior. It can be either non-associative, such as habituation or sensitization of one repeatedly occurring stimulus, or associative. For associative learning, psychology discriminates two principle forms of conditioning: operant and classical (Pavlovian). In operant conditioning, the behavior is changed in response to a comparison between an animal's own behavioral activity and its experiences (Skinner, 1938). Positive experiences tend to enforce, negative ones suppress the on-going behavior. Thus, in operant conditioning the animal exploits the consequences of its own behavior. In classical conditioning, a biologically relevant stimulus (unconditioned stimulus, US) is associated with another stimulus (conditioned stimulus, CS) occurring irrespectively of the own behavior. While the CS does normally not result in an obvious behavioral response, the US elicits an innate, often reflexive response (unconditioned response, UR). If CS and US are repeatedly paired, eventually the two stimuli become associated, resulting in a behavioral response similar to the UR, even if the previously neutral CS occurs alone. This is called a conditioned response (CR). Pioneering experiments for associative classical learning were carried out by the Russian physiologist Ivan Pavlov, who trained dogs to associate a tone (CS) with food (US) (Pavlov, 1927). The US (food) made the dogs salivate (UR), while the tone did not elicit any significant reaction before the experiment. After pairing US and CS, the dogs started to salivate as soon as they heard the sound of the bell (CR), even in the absence of food (see Figure 1).

1.2 The fruit fly *Drosophila* as a model organism

Seminal experiments on learning and memory were performed in higher organisms such as dogs (see 1.1.), monkeys or even humans. However, the possibilities to investigate the underlying genetic, molecular and cellular mechanisms are limited. For such studies, model organisms like flies, the sea slug *Aplysia* or mice are often more suitable. Especially the fruit fly *Drosophila melanogaster* has turned out to be a particularly successful model organism.

This success has many reasons. Fruit flies are conveniently small, inexpensive and easy to cultivate. Compared to the complex network of roughly 85 billion neurons in the human brain (Azevedo et al., 2009; Williams and Herrup, 1988), the function of the approximately 100000 neurons found in a fly's brain (Kei Ito, personal communication) should be easier to understand. With only four pairs of chromosomes and its complete genome being sequenced (Adams et al., 2000), a systematic analysis of *Drosophila* genetics is facilitated. Despite their relatively small size and simplicity, flies show genetic homologies with vertebrates (Rubin et al., 2000). For example, about 75% of the known human disease genes seem to have highly similar orthologues in *Drosophila* (Reiter et al., 2001). Also, the total number of genes between humans (20000 - 25000; Consortium, 2004) and *Drosophila* (about 13600; Adams et al., 2000) is comparable. Other advantages are their high fecundity and short generation time: One female can lay around 50-80 eggs per day (Novoseltsev et al., 2005) and, at room-temperature, it takes only about 10 days to develop from an egg to an adult fly (Ashburner, 1989).

Yet, the main benefit of *Drosophila*, is the availability of a variety of mutants and tools for genetic intervention (Bier, 2005; Duffy, 2002; McGuire et al., 2005). T. H. Morgan found the first *Drosophila* mutant, a white eyed fly, in 1910 (Morgan, 1910). Since then, thousands of mutants have been identified. Usually, mutagenic treatments such as the feeding of specific chemicals or UV-radiation (Ashburner, 1989) allow an unbiased generation of mutations. Another method is the use of transposable genetic elements (e.g. P-element) inserted in the flies' chromosomes (O'Kane and Gehring, 1987).

Until now, several behaviors have been described to be affected in different mutants. For example, mutations in the *period* gene discovered by Konopka and Benzer (Konopka and Benzer, 1971) affect the circadian rhythm of *Drosophila*. The discovery that mutations in this single gene could alter circadian behavior was the first step towards a molecular analysis of circadian rhythms in several species. Another example is the *fruitless* gene, which is critical for courtship behavior in *Drosophila* males (Greenspan and Ferveur, 2000). *Fruitless* mutant males do not distinguish between sexes while courting, whereas general locomotion or wing usage in males and the female behavior seem not to be altered (Goodwin et al., 2000; Villella et al., 1997). Several other mutants have been found, affecting many other behaviors, such as feeding (Fujishiro et al., 1990), vision (e.g. *optomotor blind*, (Bausenwein et al., 1986), locomotor behavior (e.g. *no-bridge*, (Strauss et al., 1992), ethanol tolerance (e.g. *tyramine β-hydroxylase*, (Scholz et al., 2000) or learning and memory.

One of the most widely used tools for genetic intervention in *Drosophila* is the GAL4/UAS system (Brand and Perrimon, 1993). This system consists of two components: a transcription factor from the yeast (GAL4) and a transgenic effector under the control of an upstream activating sequence (UAS) bound by GAL4. The two components can be combined in a simple genetic cross. In the progeny, the effector is only transcribed in those cells or tissues expressing the GAL4 protein. The GAL4 expression pattern is either mainly dependent on a known regulatory element cloned upstream to the GAL4 construct (promoter GAL4) or on the insertion site of the GAL4 construct (enhancer-trap GAL4). This expression pattern can be easily identified, for example with a green fluorescent protein (GFP) as an effector. The GAL4 expression pattern can be refined by GAL80, an inhibitor of GAL4. Furthermore, the temperature dependency of GAL4 activity provides a simple way to regulate the level of effector expression (Duffy, 2002).

Thousands of GAL4 driver lines are currently available in huge public libraries (e.g. GETDB provided by the NP consortium or a collection generated at Janelia Farm Research Campus). The separation of GAL4 and the UAS offers several advantages. First, as activation of the UAS is dependent on GAL4, the UAS construct is mostly silent in the absence of GAL4. This allows the generation of stable effector lines encoding toxic or cell ablating proteins such as tetanus toxin, diphtheria toxin or reaper (Han et al., 2000; Sweeney et al., 1995; White et al., 1996). Second, a single UAS-dependent effector can be analyzed in multiple tissues using different GAL4 drivers. Conversely, the effects of different UAS-transgenes (for example over-expression or knock-down of a specific gene) can be easily tested with the same driver line. Furthermore, tissue specific rescue experiments can be performed by restoring endogenous gene expression with appropriate drivers in a mutant background.

Despite the many advantages of the GAL4/UAS system, it also has some important limitations. First, GAL4 lines are barely specific to the cells of interest. Often additional cells are labeled. This problem can be overcome by testing different GAL4 lines with overlapping expression patterns or by introducing the GAL4 inhibitor GAL80 as a third component. Second, transcription of the UAS construct is not always absent without a driver and appears to be insertion-dependent (Ito et al., 2003; Keller et al., 2002). This potential "leakiness" could often be checked by antibody staining. Testing different insertions in different genomic regions can also help to avoid this problem.

Another important tool that can be combined with the GAL4/UAS system is the RNA interference (RNAi) technology (Figure 2). Double-stranded RNA (dsRNA) with a sequence complementary to a gene of interest is synthesized and introduced into the organism (Cerutti, 2003; Enerly et al., 2003; Giordano et al., 2002; Kalidas and Smith, 2002; Roman, 2004). The enzyme Dicer cuts this dsRNA into small interfering RNAs (siRNA). These interact with the RNA-induced silencing complex (RISC) and are cleaved into single stranded RNAs that induce the degradation of the complementary endogenous mRNA. This post-transcriptional knockdown can be made tissue specific by putting the RNAi construct under the control of UAS and driving

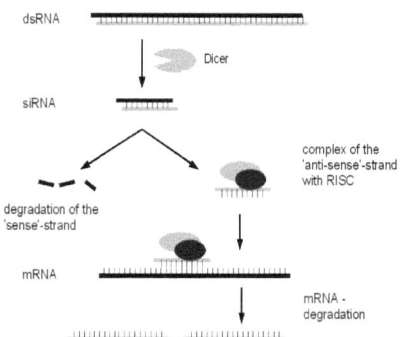

Figure 2: Mechanism of RNA interference (RNAi). The RNAi pathway is initiated by the enzyme Dicer, which cleaves long double stranded RNA (dsRNA) into 20-25 bp fragments, the small interfering RNAs (siRNA). The guiding strand of these siRNAs is incorporated in the RNA-induced silencing complex (RISC) and binds to complementary mRNA sequences of the targeted gene. The endonuclease Argonaute which is the catalytic subunit of RISC then degrades the mRNA (figure from commons.wikimedia.org).

it with appropriate GAL4 lines. A public library of UAS-RNAi constructs against more than 12000 protein coding *Drosophila* genes has been recently established (Dietzl et al., 2007). This conditional approach may circumvent the potential problem of lethality observed in some null-mutants. However, the fact that RNAi does not always lead to a complete knockout of the mRNA of the desired gene (therefore it is called "knock-down") can complicate the interpretation of the observed results. Another problem are potential off-target effects (Ma et al., 2006; Moffat et al., 2007). This can for example arise when the nucleotide sequence of the introduced RNA matches with mRNAs of other genes, and thus reduce their expression (Birmingham et al., 2006; Echeverri and Perrimon, 2006; Jackson et al , 2006; Lin et al., 2005). For this reason, software tools have been developed to improve the design of the RNAi constructs by automatically checking for cross-reactions. This risk can be further reduced by testing different constructs targeting the same mRNA. In summary, the extensive toolkit, the many practical advantages, the rich behavioral repertoire and homologies to higher vertebrates make *Drosophila melanogaster* one of the most widely used of all eukaryotic model organisms.

1.3 Associative olfactory learning and memory in *Drosophila*

Soon after the discovery of mutations affecting behavior in *Drosophila* (Benzer, 1967; Konopka and Benzer, 1971), the first mutants for learning and memory were isolated (Dudai et al., 1976; Quinn et al., 1979). With the combination of classical learning psychology and the various possibilities of genetic intervention, a new field of research was established. Odor related learning is an especially attractive branch of this field.

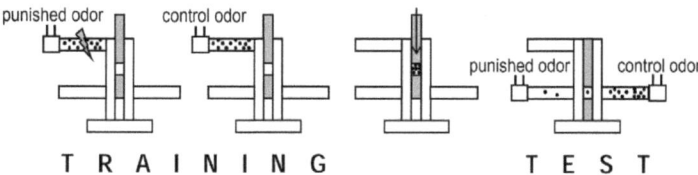

Figure 3: Paradigm for aversive olfactory learning. Air is sucked through training (top) and test tubes (bottom) in a constant flow. A group of flies (black dots) are put into the training tube covered by a copper grid, which is connected to a voltage generator. During training, flies receive one odor together with electric shocks, and a control odor without shocks. Subsequently, flies are transferred to a choice point where they can distribute between the previously punished and the control odor. Typically, flies avoid the previously punished odor (modified from Gerber et al., 2003).

The most successful paradigm for studying associative olfactory learning was introduced by Tully and Quinn (1985) and later modified by Schwärzel and colleagues (Schwaerzel et al., 2002). Briefly, a group of flies is exposed to an odor A (the CS+) which is paired with the US (e.g. electric shock punishment or sugar reward), followed by the unpaired presentation of a second odor B alone (the CS-). After this training phase, the animals are transferred to a forced choice maze, where they decide between the two olfactory cues (CS+ and CS-). Depending on the nature of the reinforcement (punishment or reward), wild-type flies typically show a significantly stronger avoidance of, or attraction to the CS+ than to the CS- during the test phase (see also Figure 3). This paradigm allows quick conditioning with a precise control over the applied stimuli, yielding a robust memory (Tully and Quinn, 1985)

In odor-shock learning, a single training session elicits memory lasting up to several hours and consists of several distinct components (Margulies et al., 2005); see also Figure 8). Short-term memory (STM) and middle-term memory (MTM) together compose the so-called anesthesia-sensitive memory (ASM) (Margulies et al., 2005). This component decays within a few hours after training. Another component, the so-called anesthesia-resistant memory (ARM) develops during the first hour after training and lasts for several hours or even days (Tully et al., 1994). These components can be genetically dissected. While MTM and ARM specifically require the *amnesiac* and the *radish* gene, respectively, *rutabaga*, *dunce* and several other genes play a preferential role in STM (Dudai et al., 1988; Folkers et al., 1993; Isabel et al., 2004; Quinn et al., 1979; Tully and Quinn, 1985).

In the setup mentioned above (Figure 3), specific forms of memory can be induced depending on the training protocol. Similar to humans and other animal models, multiple training trials with rest intervals (spaced training) have been demonstrated to be most effective in the formation of long-term memory (LTM; (Carew et al., 1972; Ebbinghaus, 1885; Frost et al., 1985; Tully et al., 1994). In *Drosophila*, this olfactory LTM lasts for several days and depends on protein-synthesis (Tully et al., 1994). In contrast, massed training (i.e. multiple training trials without any rest intervals in between) specifically elicits another form of consolidated memory (Tully et al., 1994). This memory component requires the Radish protein and is independent of protein-synthesis (i.e. it cannot be blocked by feeding of the protein-synthesis inhibitor cycloheximide).

Another advantage of the paradigm is the easily exchangeable US application (electric shock punishment or sugar reward) in the otherwise same setup. This allows a direct comparison of aversive and appetitive memory. Interestingly, sugar rewarded olfactory memory seems not to yield as high initial learning scores as the aversive, but is more stable over time (Colomb et al., 2009; Krashes and Waddell, 2008; Tempel et al., 1983). However, successful training for appetitive memory in this paradigm requires starvation of the flies (Colomb et al., 2009; Schwaerzel et al., 2003; Tempel et al., 1983; Thum, 2006)). Without starvation, they show neither a naïve nor a conditioned approach to sugar (Figure 4). This is in contrast to electric shock, where unstarved flies do react strongly to the stimulus and learn it very well, although their learning performance can also be improved by starvation (Figure 4).

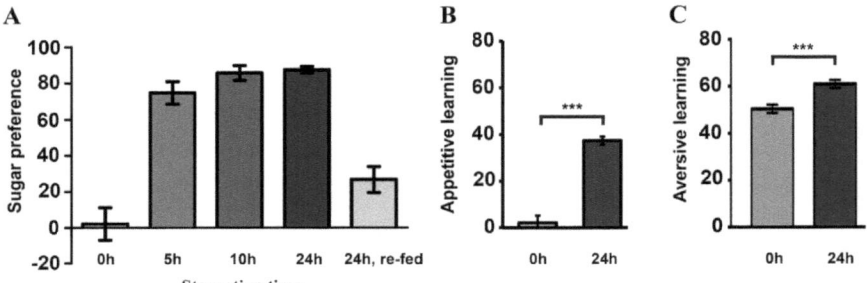

Figure 4: Starvation dependency of sugar preference and immediate memory. (A) Starved flies show a strong attraction to sugar, whereas unstarved flies do not (significance from zero $P > 0.8$). Re-feeding of 24 h starved flies for 15 minutes directly before the preference test decreases the observed attraction dramatically. (B) Unstarved flies do not show appetitive olfactory learning ($P > 0.5$; one-sample t-test against zero), whereas 24 h starved flies do. Data kindly provided by Franz Gruber. (C) Starvation is not required for aversive electric-shock learning, although memory scores are significantly increased in 24 h starved flies ($P < 0.001$).

1.4 The CS pathway: olfactory system of *Drosophila melanogaster*

The main olfactory organs of *Drosophila* are the antennae and the maxillary palps. There, olfactory input is received through sensillae that contain olfactory receptors. Up to four olfactory receptor neurons (ORN) per sensillum collect the incoming information and send their axons to the antennal lobe (AL), which functional organization is remarkably similar to that of the olfactory bulb in vertebrates (Hildebrand and Shepherd, 1997). Each individual ORN usually sends its axon to only one of the approximately 50 glomeruli in the AL. ORNs expressing the same olfactory receptor protein project to the same glomerulus (Couto et al., 2005; Fishilevich and Vosshall, 2005; Hallem and Carlson, 2004; Jefferis et al., 2001). The glomeruli are densely packed neuropils where the ORNs synapse onto local inhibitory GABAergic interneurons and about 150 projection neurons (PN) that lead to higher brain centers (Ng et al., 2002; Wilson and

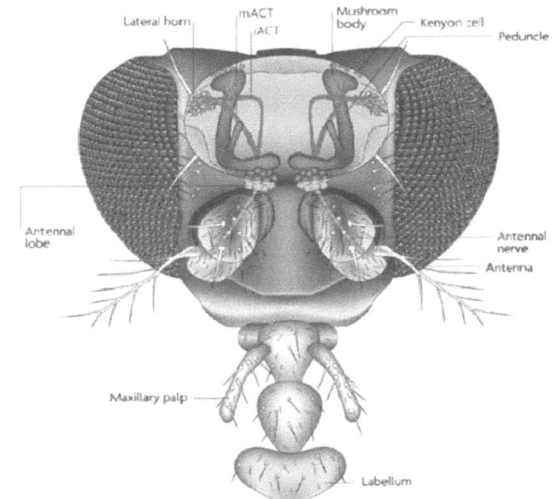

Figure 5: Olfactory system of Drosophila. Odor information is collected by olfactory receptors in the third antennal segments (about 1300 olfactory receptor neurons, ORN) and maxillary palps (about 120 ORN) and carried to the antennal lobe, where receptor fibres are sorted according to their chemospecificities in about 50 glomeruli. These represent the primary odor qualities, which are reported via the medial (mACT) and the inner antennocerebral tract (iACT) to two major target areas in the brain, the lateral protocerebrum (lateral horn) and the calyx of the mushroom body (from (Keene and Waddell, 2007).

Laurent, 2005; Yu et al., 2004). The majority of PNs receive dendritic input from only one glomerulus and send their axons via the inner antennocerebral tract (iACT) to the calyx of the mushroom body (MB) and to the lateral protocerebrum (lateral horn, (Fiala et al., 2002; Jefferis et al., 2001; Jefferis et al., 2007; Lin et al., 2007; Marin et al., 2002; Stocker et al., 1990; Tanaka et al., 2004; Wang et al., 2004; Wong et al., 2002); see Figure 5). Most of the other PNs receiving input from multiple glomeruli project directly to the lateral horn via the medial antennocerebral tract (Jefferis et al., 2002; Stocker et al., 1990). Altogether, the information gathered by the ORs undergoes first steps of processing through an interaction between the ORNs, local interneurons and PNs, before it is sent to higher brain centers (for a recent review see (Fiala, 2007).

1.5 The US pathways: The role of dopamine and octopamine for mediating punishment or reward

In insects, the biogenic amines dopamine and octopamine have been proposed to mediate punishment in aversive or reward in appetitive learning, respectively (Giurfa, 2006). In the honeybee and cricket, pharmacological blocking of dopamine receptors impaired aversive olfactory memory, whereas octopamine receptors seemed to be required for appetitive memory (Farooqui et al., 2003; Mizunami et al., 2009; Unoki et al., 2005; Vergoz et al., 2007). Injection of octopamine in the honeybee brain as well as electrical stimulation of a single octopaminergic neuron, the VUMmx1 neuron, was shown to be sufficient to substitute the reinforcing function of sucrose in an appetitive olfactory learning paradigm (Hammer, 1993; Hammer and Menzel, 1998). Similarly, in adult *Drosophila,* blocking synaptic output of dopaminergic neurons during training impairs aversive, but not appetitive olfactory learning (Schwaerzel et al., 2003). This is supported by functional imaging experiments on dopaminergic neurons innervating the MB. These neurons are strongly activated by electric shocks. Odor-induced activation was prolonged after pairing the odor with electric shock (Riemensperger et al., 2005, but see also Mao and Davis, 2009). In mutants for the *tyramine-β-hydroxylase* (*TβH*) gene encoding the enzyme for the last step of octopamine synthesis (Monastirioti et al., 1996) appetitive olfactory memory was abolished, whereas aversive memory was not significantly reduced (Schwaerzel et al., 2003). Expression of *TβH* cDNA in a set of putatively octopaminergic/tyraminergic neurons similar to the VUM cluster of the honeybee (Busch et al., 2009; Sinakevitch and Strausfeld, 2006) was sufficient to rescue this mutant learning phenotype (Thum, 2006).

A similar situation was found in *Drosophila* larvae. Substitution of punishment by artificial activation of dopaminergic neurons in an olfactory learning paradigm has been shown to be sufficient to induce aversive memory, whereas activation of tyraminergic / octopaminergic VUM (ventral unpaired medial) neurons induced appetitive memory (Schroll et al., 2006). Impairment in aversive or appetitive memory after transgenic blocking of synaptic output from dopaminergic or tyraminergic / octopaminergic neurons, respectively, supports these data (Honjo and Furukubo-Tokunaga, 2009).

However, especially the absolute selectivity of dopamine for signaling punishment in *Drosophila* is currently under debate. Kim and colleagues (Kim et al., 2007) found that adult mutants for the dopamine receptor dDA1 are strongly impaired in aversive memory, but also show some reduction in appetitive memory. Both phenotypes could be rescued by transgenic expression of a dopamine receptor in the MB (Kim et al., 2007). In addition, a recent report showed that dopamine receptor mutants are impaired in an aversive as well as in an appetitive larval learning paradigm (Selcho et al., 2009). These authors also claimed that blocking output of dopaminergic neurons affects both forms of memory in *Drosophila* larvae (Selcho et al., 2009).

Altogether, one of the fundamental roles of dopamine in *Drosophila* is signaling punishment in olfactory learning, whereas octopamine mediates the rewarding signal.

1.6 The mushroom body

The mushroom bodies (MB) are bilaterally symmetric brain structures found in many insects and other arthropods (Strausfeld et al., 1998). They were first described in 1850 by the French scientist F. Dujardin (Dujardin, 1850) and were shown to be a main center for olfactory learning and memory (Heisenberg, 2003; Menzel, 2001; Zars, 2000). The *Drosophila* MB is composed of about 2000 - 2500 major intrinsic neurons (Kenyon cells) per brain hemisphere, which accounts for more than 7% of the total brain volume (Aso et al., 2009; Rein et al., 2002; Technau and Heisenberg, 1982). The dendrites of the Kenyon cells form the calyx, which is the main olfactory input region of the MB via the PN. The cell bodies of the Kenyon cells are densely packed above and beside the calyx in the dorsocaudal cell body rind. Their axons project in a dense bundle (pedunculus) from the calyx to anterior brain regions, where they segregate into vertical (termed α and α') and medial (β, β' and γ) substructures called lobes (Crittenden et al., 1998; Ito et al., 1998), Figure 6). The α/β lobes are formed by the same neurons through bifurcation of their axons. Similarly, also the α'/β' lobes are composed by bifurcating axons, whereas in contrast the cells of the γ lobes project only to the medial lobes.

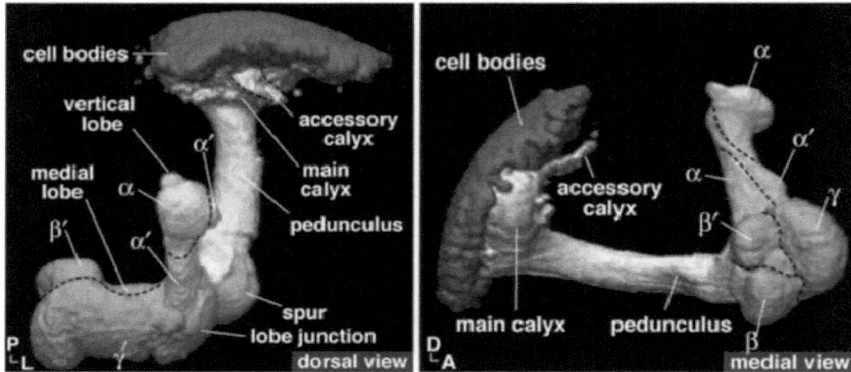

Figure 6: Organization of the mushroom body Kenyon cells. These three-dimensional (3D) reconstructions show the cell bodies (dark grey) of the Kenyon cells in the posterior cortex. Adjacently, their dendritic regions form the calyx, where the olfactory information from the projection neurons is received. From there, they extend their axons through the pedunculus (light grey) to the medial (blue; β, β' and γ) and vertical (yellow; α and α') lobes. Black dashed lines represent the border between lobes or between the main and accessory calyces. A: anterior; D: dorsal; L: lateral; M: medial; P: posterior. Scale bar = 20 µm. Adapted from (Tanaka et al., 2008).

Developmentally, the Kenyon cells derive from four neuroblasts per hemisphere (Ito et al., 1997; Ito and Hotta, 1992). Each of these neuroblasts sequentially produces the γ, α'/β' and α/β neurons (Lee et al., 1999). The earliest neurons, born between embryonic and mid-third instar larval stages, form the γ lobe. α'/β' neurons are formed between mid-third instar larval stage and puparium formation. Finally, neurons born after puparium formation compose the α/β lobes in the adult (Lee et al., 1999). In addition to their morphological distinction, the lobes can be further subdivided based on the expression pattern of different genes and enhancer trap lines, their connectivity to extrinsic neurons and their neurotransmitter systems (Johard et al., 2008; Keene and Waddell, 2007; Strausfeld et al., 2003; Tanaka et al., 2008; Yang et al., 1995).

Most of the neurotransmitters and neuromodulators implicated in MB function (e.g. dopamine, tyramine/octopamine, serotonine, acetylcholine or GABA) appear to be associated with extrinsic neurons (Davis, 2005; Heisenberg, 2003; Kim et al., 2007; Schroll et al., 2006; Schwaerzel et al., 2003; Waddell et al., 2000; Zhang et al., 2007), whereas little is known about the neurotransmitters functioning within the Kenyon cells. A small set of Kenyon cells has been shown to be glutamatergic (Strausfeld et al., 2003), but so far it is not clear whether this is a metabolic intermediate or whether it has actual neurotransmitter function. Until now, the most promising candidate for a functional Kenyon cell neurotransmitter is short neuropeptide F (sNPF), which has been demonstrated recently to be expressed in most of the MB Kenyon cells (Johard et al., 2008).

The requirement of the MB for olfactory learning and memory was first demonstrated in two structural brain mutants with deranged or reduced MBs (Heisenberg et al., 1985). Similarly, chemical ablation of the MB with hydroxyurea also caused impaired olfactory learning (de Belle and Heisenberg, 1994). Transgenic overexpression of a constitutively active $G_{\alpha s}$ protein subunit in the MB by means of the GAL4/UAS system completely blocked olfactory aversive learning (Connolly et al., 1996)). A temperature-sensitive, dominant-negative allele of the *Drosophila* dynamin gene *shibire* (shi^{ts1}) under control of UAS (Kitamoto, 2001) allowed conditional blocking of chemical output at the synapse and revealed that MB output is required only during retrieval of memory, but not during acquisition (Dubnau et al., 2001; McGuire et al., 2001; Schwaerzel et al., 2002). Furthermore, the memory impairment of *rutabaga* (*rut*) mutants can be rescued by restoring *rut* expression in MB neurons at the adult stage (Mao et al., 2004; McGuire et al., 2003; Zars et al., 2000). Thus, the MB is necessary and sufficient for associative plasticity underlying olfactory aversive memory and Kenyon cells are the most likely candidates to house the memory trace for the association between odor and shock (Figure 7, reviewed in Gerber et al., 2004).

However, the different subtypes of Kenyon cells can also be dissociated on the behavioral level. The α/β and especially the γ lobes have been suggested to be required for STM, because rescue of the *rut* mutant phenotype in STM was achieved by restoring *rut* expression in the γ lobes (Blum et al., 2009; Zars et al., 2000) or, to a full extent, in all of these lobes (Akalal et al., 2006; McGuire et al., 2003). In addition, blocking synaptic output from the α/β lobes with shi^{ts} impaired STM (Akalal et al., 2006; McGuire et al., 2001) as well as expression of the constitutively active $G_{\alpha s}$ subunit in the γ lobes (Connolly et al., 1996). During acquisition and consolidation of memory measured at 2-3 hours after training, output from the α'/β' lobes as well as from the connected DPM neurons seems to be required (Krashes et al., 2007). Retrieval of this memory depends on output from the α/β lobes (Krashes et al., 2007; McGuire et al., 2001). Thus, the memory trace for 2-3 hour memory should be in the α/β lobes. However, the γ lobes may also be involved in this memory phase (Cheng et al., 2001; Isabel et al., 2004). In contrast to that, for ARM only the α/β lobes seem to be required, whereas blocking of synaptic output from the γ lobes during training and test seems to have no effect (Isabel et al., 2004). Finally, LTM might preferentially depend on the α lobe (Blum et al., 2009; Isabel et al., 2004; Pascual and Preat, 2001; Yu et al., 2006).

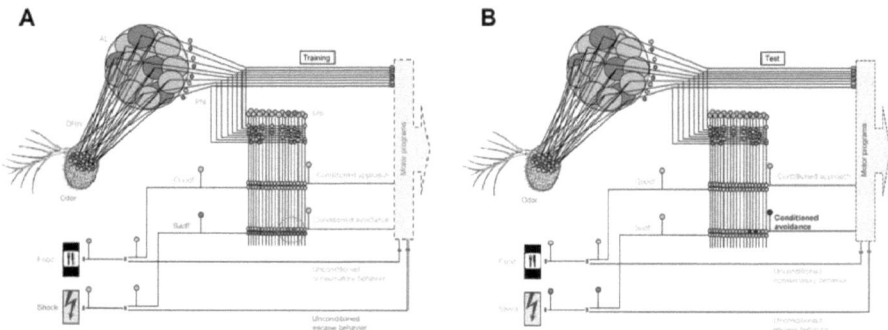

Figure 7: A minimal model for the role of mushroom body Kenyon cells in *Drosophila* olfactory learning. A highly simplified diagram shows the olfactory pathway. Olfactory receptor neurons (ORN) project to the antennal lobe (AL), leading to a specific combinatorial activity pattern. From there, uniglomerular projection neurons (PN) relay to the lateral horn and to premotor centers (box labeled 'Motor output'), as well as to the mushroom body (MB) calyx. Output from the MB then projects to a variety of target regions including premotor areas. In this model, we assume that a Kenyon cell needs input from at least three projection neurons to fire. In the mushroom bodies, the activation pattern of the sensory and the projection neurons is transformed into an activation pattern of the MB-intrinsic Kenyon cells. A memory trace for the association between odor and reinforcement is proposed to be localized within the Kenyon cells: when during training the activation of a pattern of Kenyon cells representing an odor occurs simultaneously with a modulatory reinforcement signal (labeled 'Good!' and 'Bad!'; potentially octopamine and dopamine signaling, respectively), output from these activated Kenyon cells onto MB output neurons is suggested to be strengthened. This strengthened output is thought to mediate conditioned behavior towards the odor when encountered during test, during which no reinforcer is present. Activated cells or synapses and motor programs are represented by filled symbols and bold lettering, respectively. (A) situation during training, (B) test. Taken from (Gerber et al., 2004).

In the model shown in Figure 7, activation of a pattern of Kenyon cells representing an odor occurs simultaneously with a modulatory reinforcement signal. This strengthens the synaptic output of the activated Kenyon cells onto MB output neurons and is thought to mediate the conditioned behavior (Heisenberg, 2003). Similarly as for aversive odor-memory, the MB has also been shown to be necessary and sufficient for appetitive learning (Schwaerzel et al., 2003). However, besides from the MB, *rut* rescue was also shown to be sufficient in the projection neurons (Thum et al., 2007), suggesting another redundant memory trace for appetitive learning there.

1.7 Different components of aversive olfactory memory

Similar to memory in vertebrates, associative olfactory memory of *Drosophila* is dynamic (McGaugh, 2000). Already after a single training session, flies form an aversive memory that lasts for several hours and consists of qualitatively different components (Figure 8). These components can be dissociated by their susceptibility to amnesic treatments such as cold-induced anesthesia as well as by their underlying cellular and molecular circuits (Dubnau and Tully, 1998; Isabel et al., 2004; Keene and Waddell, 2007; Masek and Heisenberg, 2008; Xia and Tully, 2007).

A relatively quickly decaying short-term memory (STM) can be disrupted by cold-induced anesthesia directly after training (Quinn and Dudai, 1976; Tully et al., 1994; Figure 8). For this learning/STM component several genes such as *latheo, linotte, 14-3-3 (leonardo), scabrous (volado), fasII* and *DC0* (PKA) have been found to be preferentially required (Boynton and Tully, 1992; Cheng et al., 2001; Dura et al., 1993; Grotewiel et al., 1998; Skoulakis and Davis, 1996; Skoulakis et al., 1993). While the STM decays quickly after training, another anesthesia-sensitive memory component, the middle-term memory (MTM), lasts significantly longer. MTM is specifically impaired in a mutant for the *amnesiac* (*amn*) gene, which putatively encodes a homologue of the mammalian pituitary adenylate cyclase-activating peptide (PACAP, Feany and Quinn, 1995; Moore et al., 1998). *amn* is supposed to function in two exclusively

MB-projecting neurons, the DPM neurons, for stabilizing both aversive and appetitive odor-memory (Feany and Quinn, 1995; Keene et al., 2006; Keene et al., 2004; Moore et al., 1998; Quinn et al., 1979; Tamura et al., 2003; Waddell et al., 2000; Yu et al., 2005). Flies that express an endogenous competitor of the PKA-RII subunit binding PKA to A-kinase anchoring proteins (AKAPs) in the mushroom body have a defect in anesthesia-sensitive memory component (ASM), whereas ARM is unaffected (Feany and Quinn, 1995; Schwaerzel et al., 2007). In both cases, initial learning seems to be intact. Together, STM and MTM compose the labile ASM.

While mutants for the cAMP/PKA signaling pathway like *amn* or *rut* are selectively impaired in ASM, the *radish* (*rad*) gene is specifically required for another memory component, the anesthesia-resistant memory (ARM) (Dudai et al., 1988; Folkers et al., 1993; Isabel et al., 2004; Quinn and Dudai, 1976; Schwaerzel et al., 2007; Tully et al., 1994). This consolidated component is built gradually within the following hours after training and can last up to days (Figure 8, Margulies et al., 2005; Quinn and Dudai, 1976). Other than by single training, it can be specifically elicited by the application of a massed training protocol (Tully et al., 1994).

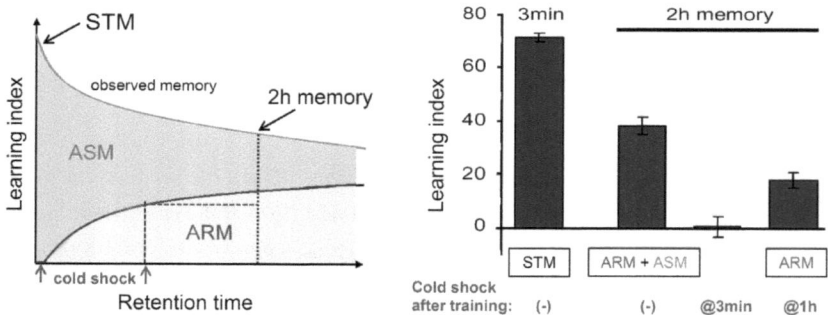

Figure 8: Memory components after single training. Immediate olfactory memory (short-term memory, STM) measured at three minutes after training is susceptible to amnesic treatments such as cold shock. Although this observed memory after a single training session decays over time, even after several hours flies show still a significant memory score. However, this memory is not uniform. With time, a consolidated anesthesia-resistant memory component (ARM) forms, while the anesthesia-sensitive components (ASM) decay. This model can be tested experimentally (right panel).

In contrast, another consolidated memory component, the protein-synthesis dependent long-term memory (LTM), forms only after spaced training and is intact in the *rad* mutant (Tully et al., 1994). LTM has been shown to specifically require the cAMP response element binding protein (dCREB) as well as other transcription factors like Adf1 (*nalyot*) and Notch. Further genes for LTM are *nebula, tequila, cramer* and the *staufen/pumilio* pathway. Also dNR1, the *Drosophila* homologue of the postsynaptic mammalian NMDAR1 receptor, has recently been shown to be involved in olfactory learning and subsequent LTM formation (Xia et al., 2005). Interestingly, several proteins encoded by the afore-mentioned genes like Notch, Cramer or Nebula are supposed to have human homologues putatively involved in mental diseases, such as Alzheimer's or the Down's, Alagille and Cadasil syndromes.

1.8 Molecular mechanisms of olfactory learning

The *rut* dependent cAMP/PKA signaling pathway has been suggested to play a critical role for olfactory learning (Figure 9). GTP-binding protein (G protein)-coupled receptors for dopamine and octopamine have been found to be involved in the regulation of cAMP levels (Balfanz et al., 2005; Han et al., 1998; Han et al., 1996; Hearn et al., 2002; Kim et al., 2003; Lee et al., 2003; Maqueira et al., 2005). Both receptor types as well as a *rut*-dependent adenylyl cyclase (AC), are highly expressed in the mushroom body (Han et al., 1998; Han et al., 1996; Han et al., 1992; Levin et al., 1992), which is supposed to be the major site of olfactory memory formation in *Drosophila* (Gerber et al., 2004; see previous chapter). Synthesis of cAMP by this AC, which is homologous to the mammalian type-1 adenylyl cyclase, is activated by G-protein- and Ca^{2+}/Calmodulin-dependent stimulation (Dudai et al., 1988). Interestingly, the two forms of stimulation seem to act synergistically, such that Ca^{2+}/Calmodulin stimulation along with G-protein stimulation leads to a more than additive activation of the AC than either stimulus alone. Therefore, this cyclase is thought to act as a molecular coincidence detector between the conditioned stimulus (odor) and the reinforcer during classical learning (Abrams and Kandel, 1988; Anholt, 1994; Dudai et al., 1988).

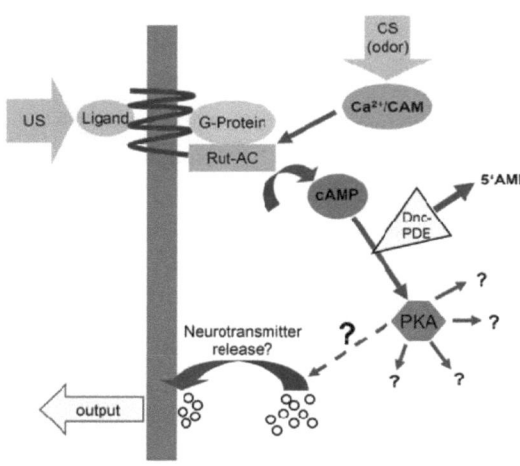

Figure 9: Molecular model of olfactory learning. Presynaptic modulation of transmission at the Kenyon cell synapse to the output neuron is thought to underlie olfactory short and middle term memory in *Drosophila*. Arrival of the CS and US activates the adenylyl cyclase A (Rut-AC) via calcium/calmodulin (Ca^{2+}/CAM) increase and G-protein coupled receptors, respectively. This leads to the synthesis of cAMP. If these activations of the Rut-AC take place simultaneously, cAMP levels are synergistically increased, which suggests the cyclase as a molecular coincidence detector. cAMP activates the protein kinase A (PKA), which in turn may phosphorylate synaptic proteins, thus evoking neurotransmitter release. A Phosphodiesterase (Dnc-PDE) regulates cAMP levels by degrading cAMP to 5'AMP.

The importance of the cAMP signaling cascade for olfactory memory is well established. First, *rut* mutants (rut^{2080}, rut^1) deficient in the AC itself are severely impaired in their learning ability (Davis et al., 1995; Thum, 2006; Tully and Quinn, 1985). By expressing the *rut* gene exclusively in subsets of the mushroom body Kenyon cells in otherwise *rut*-deficient flies (Akalal et al., 2006; Mao et al., 2004; Zars et al., 2000) it was shown that the AC is not only necessary but also sufficient in these neurons for olfactory aversive learning. Second, a mutant for *dunce* (*dnc*), encoding a phosphodiesterase degrading cAMP (Byers et al., 1981), is impaired in learning. As in the case of Rut, the Dnc protein is also preferentially expressed in the mushroom body (Nighorn et al., 1991). Third, olfactory learning is abolished if the regulation of the cAMP signaling is disrupted by a constitutively active $G\alpha_s$ subunit in the mushroom body (Connolly et al., 1996). In addition, expression of the catalytic and regulatory subunits of the protein kinase A (PKA), which is the major downstream effector of cAMP (Levin et al., 1992), is elevated in the mushroom body (Crittenden et al., 1998; Skoulakis et al., 1993). PKA mutants or inducible inhibitory transgenes for PKA also show deficiencies in learning and memory (Drain et al., 1991; Goodwin et al., 1997; Li et al., 1996; Skoulakis et al., 1993).

1.9 Regulation of presynaptic neurotransmitter release

Synaptic plasticity is thought to provide the structural basis underlying learning and memory. As chemical synapses pass information directionally from a presynaptic to a postsynaptic cell they are asymmetric in structure and function. The end of a neuron's axon has branching terminals that release neurotransmitters into a gap called the synaptic cleft between the terminals and the dendrites of the next neuron. This neurotransmitter release is triggered by Ca^{2+} influx after an incoming action potential. The released neurotransmitters bind to receptors (e.g. the glutamatergic N-methyl-D-aspartate [NMDA] or α-amino-3-hydroxy-5-methyl-4-isoxazolepropionic acid receptor [AMPA] receptor) at the postsynaptic neurons, which in turn open for Ca^{2+} and other cations, thereby allowing transmission of the signal to the next cell.

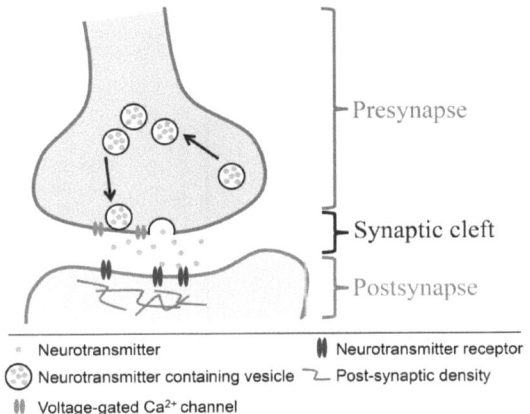

Figure 10: Simplified schematic of a synapse. Chemical synapses pass information directionally from a presynaptic to a postsynaptic cell and are therefore asymmetric in structure and function. The presynaptic terminals of an axon house neurotransmitter containing vesicles, which are released at the active zone (AZ) into the synaptic cleft. Immediately opposite of the AZ lies a region of the postsynaptic cell called postsynaptic density (PSD). The PSDs are composed of neurotransmitter receptors and an elaborate complex of interlinked proteins that are involved in anchoring, trafficking and modulation of the activity of these receptors. The membranes of the two adjacent cells are held together by cell adhesion proteins. The small size of the cleft (about 20nm) allows a tight regulation of the neurotransmitter concentration.

Regulation of presynaptic neurotransmitter release is supposed to be a prerequisite for synaptic and therefore behavioral plasticity. At the presynapse, neurotransmitter-containing vesicles are organized into at least two distinct pools with different release probabilities (Pieribone et al., 1995). While a relatively small and quickly exhausted readily releasable pool of vesicles is located directly at the active zones, the majority of vesicles are stored in a reserve/recycling pool distant from the actual release sites (Hilfiker et al., 1999; Figure 11). The highly abundant and phylogenetically conserved Synapsin (Syn) proteins seem to be specifically required for the reserve pool (Hilfiker et al., 2005; Hilfiker et al., 1999). A current working model assumes that Synapsin tethers vesicles together and to the cytoskeleton (Hilfiker et al., 1999; but see also Gaffield and Betz, 2007) for a critical discussion). Release from this reserve pool is triggered by the phosphorylation of Synapsin. Thus, the phosphorylation status of Synapsin regulates the number of vesicles available for release (Chi et al., 2001; Gitler et al., 2008; Hosaka and Sudhof, 1999; Menegon et al., 2006). If the function of Synapsin is compromised, neurotransmitter release specifically under sustained high-frequency stimulation is impaired, whereas synaptic output *per se* is still possible (Chi et al., 2003; Gitler et al., 2004). In

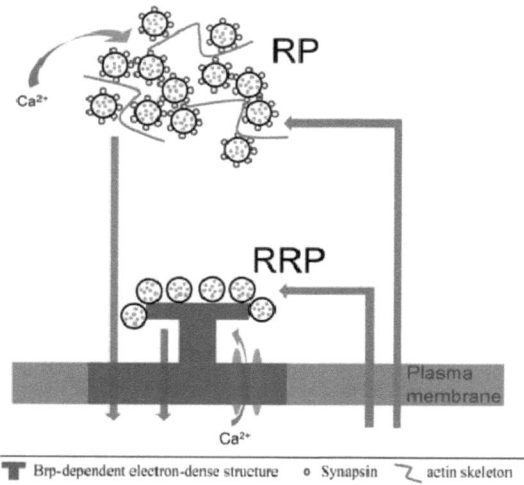

Figure 11: Vesicle pool model at the presynapse. The vesicle protein Synapsin is supposed to tether the vesicles together and to the actin skeleton, thereby building a reserve pool (RP) of vesicles. Upon phosphorylation of Synapsin, induced by extracellular Ca^{2+} influx, vesicles can travel to the plasma membrane and release their neurotransmitter contents. Vesicle recycling to the RP has been shown to require cAMP. A smaller readily-releasable pool (RRP) is located directly at the active zones. Vesicles are docked to Bruchpilot (Brp)-dependent electron-dense structures (T-bars). Local Ca^{2+} influx through cacophony-dependent channels in the vicinity of the T-bars triggers fast neurotransmitter release.

contrast to vertebrates where three *synapsin* (*syn*) genes are found, the *Drosophila* genome contains only a single one (Hilfiker et al., 1999; Klagges et al., 1996). Almost all Synapsins investigated so far share domains A, C and E, pointing to a conserved function mediated by these domains (Kao et al., 1999). Recently, the *Drosophila syn^{97}* allele was described as carrying a 1.4-kb deletion spanning parts of the regulatory sequence of the *syn* gene and half of its first exon. As a consequence, adult *syn^{97}* mutants lack detectable Synapsin levels and therefore qualify as null mutants (Godenschwege et al., 2004).

One of the key proteins for the formation of the readily releasable pool in *Drosophila* seems to be Bruchpilot (Brp), which is homologous to the mammalian ELKS/CAST proteins (Wagh et al., 2006). At the *Drosophila* neuromuscular junction, Brp-knock-down specifically disturbs the accumulation of synaptic vesicles directly at the release sites, because the electron-

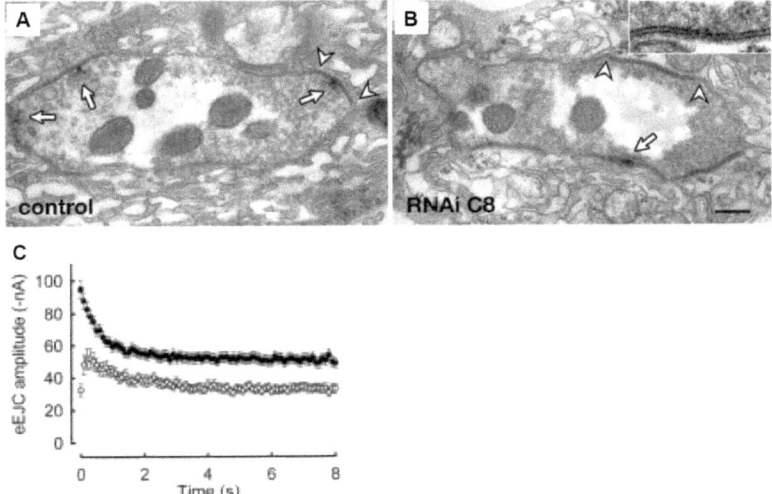

Figure 12: Electron microscopy (EM) and electrophysiology at NMJs of Brp-impaired larvae and controls. (A) In controls, individual synaptic densities (example marked by arrowhead) at larval neuromuscular junctions are often decorated with a pre- synaptic T-bar (arrows). (B) When Brp is knocked down, normally appearing synaptic densities (inset) are formed, but mostly lack T-bars (arrow). (C) Synaptic transmission of *brp* mutants after 10 Hz electrical stimulation is most significantly affected in comparison to WT controls (black circles) at the very first pulse, whereas sustained transmitter release is still possible. (A) and (B) from Wagh et al., 2006, (C) from Kittel et al., 2006.

dense structures (T-bars) required for the docking of vesicles at the active zone are severely impaired (Figure 12; Kittel et al., 2006; Wagh et al., 2006). The Bruchpilot protein itself is probably an essential component of the T-bars, with its N-terminal sticking to the plasma membrane and the C-terminus (detected by the monoclonal antibody nc82) forming the vesicle docking site of the T-bars (Fouquet et al., 2009). The Brp N-terminals are suggested to physically interact with the *cacophony*-dependent voltage-gated Ca^{2+} channels normally found in direct vicinity of the T-bars (Fouquet et al., 2009). This is consistent with the fact that these Ca^{2+} channels are dispersed in *brp* mutants (Kittel et al., 2006). As specifically the very immediate neurotransmitter release is most significantly affected in *brp* null-mutants (Figure 12C), this supports a selective function of Brp for immediate vesicle release (Fouquet et al., 2009; Kittel et al., 2006).

1.10 Aim of the work

Although our knowledge about the molecular processes underlying olfactory learning and different memory components in *Drosophila* is constantly growing, hardly anything is known about their synaptic mechanisms and whether distinct forms of neurotransmission dissociate different memory components. Therefore, the main aim of this work is to find specific synaptic correlates for ASM and ARM. To this end, I chose the presynaptic proteins Synapsin and Bruchpilot, which are supposed to be required for distinct forms of neurotransmitter release, and tested their role in different components of olfactory memory.

2 Material and Methods

2.1 Fly care and genotypes

Flies were raised at 25 °C and 60 % humidity with a 14 h : 10 h (Würzburg) or a 12 h : 12 h (Martinsried) light : dark cycle on standard corn-meal based food. Different conditions between Würzburg and Martinsried seemed not to critically alter the investigated behaviors. For behavioral experiments, a null mutant for *synapsin* (CG3985, *synapsin^{97}* [*syn^{97}*]) which had been backcrossed 13 times to wild-type Canton-S (Michels et al., 2005) and the respective wild-type Canton-S flies were used. For the *synapsin* double mutant experiments in Figure 19A, a homozygous hypomorphic mutant for *rutabaga* (CG9533, *rut^{2080}*; Pan et al., 2009) and for *dunce* (CG32498, *dnc^{1}*; Nighorn et al., 1991), the *rut^{2080}*; *syn^{97}* and the *dnc^{1}*; *syn^{97}* double mutants (this study) have been tested. For the experiments in Figure 19B female homozygous mutants for *rut^{2080}*, *dnc^{1}*, *rut^{2080}*; *syn^{97}* or *dnc^{1}*; *syn^{97}* were crossed to CS or *syn^{97}* males. There, a mixture of male and female progeny was tested in the olfactory conditioning, but only the male progeny was used to calculate the learning indices.

For the Bruchpilot experiments, besides the wild type Canton-S and the *rut^{2080}* mutant, the F1 progeny of respective crosses were used. As effector lines, three homozygous UAS-*bruchpilot*-RNAi lines (*w$^-$*; RNAi^{B3} (III); Wagh et al., 2006, *w$^-$*; RNAi^{C8} (III); Wagh et al., 2006 and *w$^-$*; RNAi^{B3,C8} (III), combination of RNAi^{B3} and RNAi^{C8}, this study) were used. GAL4 driver lines employed were the homozygous *w$^-$; OK107* (IV) (Connolly et al., 1996), *w$^-$; MB247* (III) (Zars et al., 2000) and *w$^-$; c739* (II) (Yang et al., 1995). To generate the genetic controls, either the GAL4 or the UAS lines were crossed to *w$^-$* flies. For the experiment in Figure 23B, flies carrying the GAL4-repressor GAL80 (Kitamoto, 2002) under control of the MB247 promoter (*w$^-$; MB247*-GAL80 (II) (Krashes et al., 2007) were recombined with the UAS-RNAi^{B3,C8} construct. These *w$^-$; MB247*-GAL80; RNAi^{B3,C8} flies were crossed to either *w$^-$; OK107* or *w$^-$; c739*. In Figure 25, female flies carrying UAS-RNAi^{B3,C8} in a *rut^{2080}* (X) mutant background (*rut^{2080}*; RNAi^{B3,C8}) were crossed to *w$^-$; OK107* or *w$^-$* males. To ensure the knockdown with *bruchpilot*-

RNAi, all vials were shifted from 25°C to 29°C when the larvae reached late 3rd instar state for these experiments.

All behavioral experiments were performed at room temperature (22 ± 3 °C) with relative humidity at 80 ± 5 %. One- to six-day old adult flies were trained under dim red light, which prevents the flies' vision, and tested in complete darkness. All flies were collected and transferred to fresh food vials at least 24 h before the experiments. Mixed populations of males and females were measured and used to calculate the performance indices, except for the experiments in Figure 19B and Figure 25, where the sexes have been separately counted after the test and only males were used for calculation due to the crossing regimen (see above).

Table 1: Employed fly strains

Line	Description	Comments	Reference
Canton Special (CS)	wild type	from Würzburg stock collection	(Schwaerzel et al., 2002)
***white*^{1118}**	*white*$^-$ mutant	cantonized	(Dura et al., 1993)
***syn*^{97}**	*synapsin* mutant, on 3rd chromosome	null mutant, cantonized	(Godenschwege et al., 2004; Michels et al., 2005)
***rut*^{2080}**	*rutabaga* mutant, P-element containing *rosy*$^+$ and *lacZ* cDNA, on X chromosome	hypomorph, P-element induced mutation, cantonized, insertion checked with XGal staining	(Levin et al., 1992)
***rut*^{2080}; *syn*^{97}**	*rutabaga*; *synapsin* double mutant, on X and 3rd chromosome		this study
***dnc*^{1}**	*dunce* mutant, on X chromosome	hypomorph	(Byers et al., 1981)

dnc^1; syn^{97}	*dunce*; *synapsin* double mutant, on X and 3rd chromosome		this study
UAS-*bruchpilot*-RNAi^{B3}	*white*⁻, P-element containing *white*⁺ cDNA, on 3rd chromosome	targets to CG30336 region of the *brp* gene	(Wagh et al., 2006)
UAS-*bruchpilot*-RNAi^{C8}	*white*⁻, P-element containing *white*⁺ cDNA, on 3rd chromosome	targets to CG30337 region of the *brp* gene	(Wagh et al., 2006)
UAS-*bruchpilot*-RNAi^{B3C8}	*white*⁻, P-elements containing *white*⁺ cDNA, on 3rd chromosome	bears B3 and C8 construct	Stephan Sigrist
OK107-GAL4	*white*⁻, P-element containing *white*⁺ cDNA, on 4th chromosome	MB-GAL4, P-element inserted near *eyeless* gene	(Connolly et al., 1996)
MB247-GAL4	*white*⁻, P-element containing *white*⁺ cDNA, on 3rd chromosome	MB-GAL4, expression controlled by regulatory region of the *D-Mef2* gene	(Zars et al., 2000)
c739-GAL4	*white*⁻, P-element containing *white*⁺ cDNA, on 2nd chromosome	MB-GAL4, preferentially in α/β	(Yang et al., 1995)
247-GAL80, UAS-*brp*-RNAi^{B3C8}	*white*⁻, P-elements containing *white*⁺ cDNA, on 3rd chromosome	UAS-RNAi expression blocked in the MB by GAL80	this study
rut^{2080}; UAS-*brp*-RNAi^{B3C8}	rut^{2080} mutant background, *brp*-RNAi^{B3C8} on 3rd chromosome		this study

2.2 Immunohistochemistry

For the Synapsin experiments presented in Figure 15, brains of 4-6 day-old female flies were prepared and fixed in 2% formaldehyde for 1h at room-temperature. Rabbit polyclonal anti-GFP (Molecular Probes, Eugene, OR ; 1:2000) and mouse monoclonal anti-Synapsin, 3C11 (1:20) (Klagges et al., 1996) were used as primary antibodies. Alexa488-conjugated goat anti-rabbit antibody (Molecular Probes, Eugene, OR; 1:1000) and Cy3-conjugated goat anti-mouse antibody (Jackson Immunoresearch, West Grove, PA; 1:250) were applied as secondary antibodies. Confocal image stacks of whole mount brains were taken with a Leica SP1. A stack of images was collected at 1 µm steps with a ×20 objective (Figure 15A, E) and further magnified with a ×40 objective at a step size of 0.7 µm. Images of the confocal stacks were projected and analyzed with the software Image-J (NIH). The adjustments of contrast and brightness as well as rotations of the images were done with Photoshop (Adobe, San Jose, CA).

For the Brp knockdown experiments shown in Figure 20 brains of female flies were prepared and fixed in 2 % paraformaldehyde for 1h at room-temperature. Mouse monoclonal anti-Brp, nc82 (Hofbauer et al., 2009), and Alexa488-conjugated goat anti-mouse antibody () were applied as the primary and secondary antibodies, respectively. Brains for Brp quantification were counterstained with Phalloidin (diluted 1:300, Molecular Probes; Eugene, OR). Confocal image stacks of whole mount brains were taken with a Leica SP1 or Olympus FV1000. Projection of the confocal stacks and quantification signal intensity were done with ImageJ (NIH). The adjustments of brightness/contrast were done with Photoshop (Adobe, San Jose, CA).

Quantification of the Brp signal was done by selecting a confocal slice containing the horizontal lobes of the mushroom body from a stack. The average signal intensity of a pixel in the experimental (the gamma lobe of the mushroom body) and the neighboring control region (anterior superior medial protocerebrum) was measured by "Histogram" function in ImageJ. The typical size of the region of interest was approximately 1000-2500 pixels. The signal of the experimental region was normalized by the control region.

2.3 Olfactory conditioning

Standard differential olfactory conditioning was performed (Schwaerzel et al., 2002; Tully and Quinn, 1985). Approximately 100 flies were exposed alternately to two different odors. With one of these odors, 12 pulses (1.4 s duration each with onset-onset interval of 5 s) of electric shock were contingently applied. In a subsequent test in a T-maze, both punished and control odors were presented simultaneously; flies were then allowed to choose between them during the subsequent 100 seconds (120 seconds in case of all Bruchpilot experiments). To correct for a learning-independent preference to one of the two odors, two separate groups of flies were trained reciprocally: one group received odor A with shock (A+) and B without (B-), and the other group received odor B with shock (B+) and A without (A-). A learning index (LI) was then calculated as the mean preference of these two reciprocally trained groups (N = the number of flies running to the indicated odor):

$$LI = [\{(N_{A-} - N_{B+}) / (N_{A-} + N_{B+}) + (N_{B-} - N_{A+}) / (N_{B-} + N_{A+})\} \times 100] / 2$$

To rule out non-associative effects caused by the order of the reinforcement, in half of the experiments the first presented odor was punished (A+, B- and B+, A-) and *vice versa* in the other half (A-, B+ and B-, A+). When not tested immediately after training, flies were kept in small vials between training and test).

2.4 Anesthesia-resistant memory (ARM)

To measure ARM, flies were transferred immediately after training into empty plastic vials, which were put in a light-proof box and stored at room temperature. At a respective time point after training (see Legends), these vials were put into an ice-floating water bath, which anesthetized all flies within about 20 s. After two minutes, the flies were transferred to prewarmed vials and kept at room temperature until the test. They typically recovered from the anesthesia within about 30 s.

2.5 Responsiveness to electric shocks and odors

Flies were tested for responsiveness to electric shock and the odors in the same T-maze assay as used for the learning experiments. For electric shock responsiveness, flies were given one minute to choose between an electrified (90V, 12 pulses as above) and a non-electrified arm of the T-maze. From each experiment, the number of flies choosing the electrified (N_+) or the non-electrified arm (N_-) was counted and a performance (avoidance) index (PI) was calculated as:

$$PI = \{(N_- - N_+) / (N_- + N_+)\} \times 100$$

To test the responsiveness to olfactory cues, flies were given 2 minutes to choose between two arms of the T-maze; one scented with the respective odor used for conditioning and the other one unscented. A PI was calculated from the number of flies choosing either the scented (N_+) or the unscented arm (N_-):

$$PI = \{(N_- - N_+) / (N_- + N_+)\} \times 100$$

Synapsin mutants have been reported to show a habituation phenotype (Godenschwege et al., 2004). To test for mutant defects entailed by the mere exposure to either odors or electric shocks in these flies, odor preference was additionally tested after treatments where either only odors or only shocks were administered in the same regimen as during conditioning (sham-training) (Michels et al., 2005).

2.6 Employed odors

As odors, either the undiluted chemicals benzaldehyde (Fluka, Munich, Germany) and (R)-(+)-limonen (Sigma Aldrich, Munich, Germany) presented in cups with a diameter of 5 mm and 7 mm, respectively, or 4-methylcyclohexanol (Fluka, Munich, Germany, diluted 1: 80) and 3-octanol (Fluka, Munich, Germany, diluted 1:100) in 14mm odor cups were used as specified in the legends. Bruchpilot experiments employed 4-methylcyclohexanol (Fluka, Munich, Germany, diluted 1:80) and 3-octanol (Fluka, Munich, Germany, diluted 1:100) in odor cups with a diameter of 14 mm. For Figure 25, benzaldehyde (VWR International GmbH, Darmstadt, Germany) diluted 1:1000 in a 16mm cup was used instead of 4-methylcyclohexanol. If applicable, paraffin oil (Fluka, Munich, Germany) was used for dilution.

2.7 Statistics

Significance level in each experiment was set to 5%. The data were first tested for normal distribution and homogeneity of variance with Shapiro-Wilk test followed by Bonferroni correction and Bartlett's test, respectively. If none of these assumptions were violated, parametric comparisons (one-way ANOVA followed by Bonferroni-corrected *post-hoc* tests or Pearson product-moment correlation coefficient) were applied. If data were not normally distributed or homogeneity of variances was violated, non-parametric comparisons (Kruskal-Wallis test followed by Dunns correction) were performed. All statistical calculation were done using the software Prism (GraphPad, San Diego, CA).

3 Results

3.1 Reinforcer intensity differentially affects distinct memory components

One of the critical factors that determine the magnitude of aversive olfactory memory is the intensity of reinforcement (Diegelmann et al., 2006b; Dudai et al., 1988; Tempel et al., 1983; Tully and Quinn, 1985). The stronger the reinforcement, the better the learning performance. This prompted me to ask whether the different reinforcer intensities can induce qualitatively distinct types of memory. For this purpose olfactory conditioning in *Drosophila* is well suited, because the reinforcer can be controlled by adjusting the voltage of the electric shocks. The higher the shock intensity of 12 consecutive electric shock pulses, the higher the naïve shock avoidance of wild type flies (Figure 13A). I first examined immediate (~ 3 min) and 5-hour memories after a single training session at various intensities (Figure 13B). Immediate memory, as mentioned above, followed the strength of electric shocks; so did the 5-hour memory (Figure 13B). Conditioned odor avoidance tested immediately and 5 hours after training with different shock intensities were both linearly related to the naïve shock avoidance (Figure 13C; $P = 0.0019$ for correlation between naïve avoidance and immediate memory; $P = 0.0217$ for correlation between naïve avoidance and 5h memory; Pearson's r test).

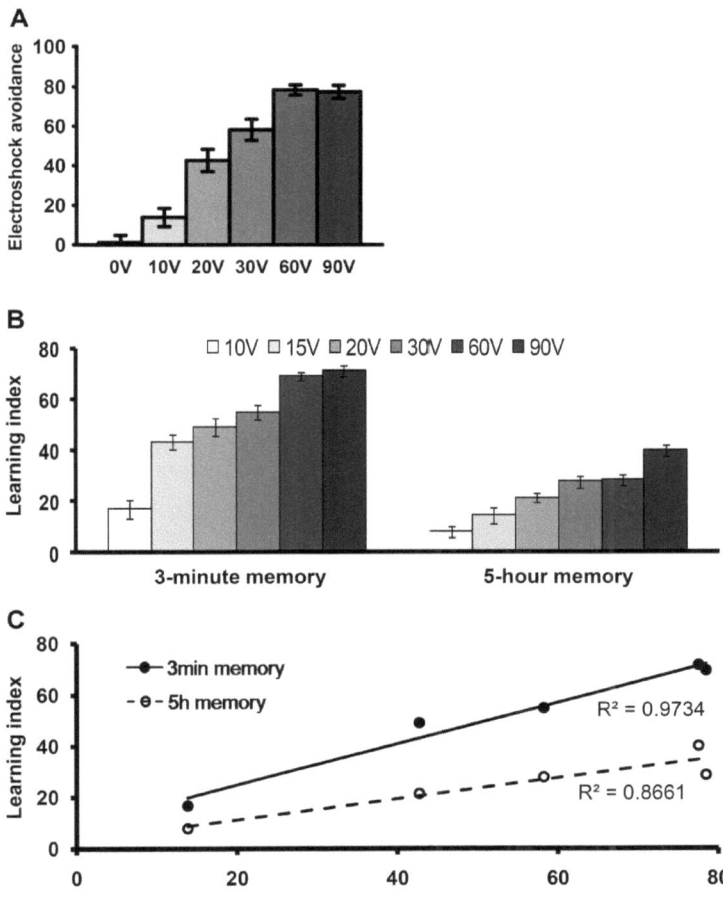

Figure 13: Olfactory memories and their dependency on reinforcer intensity. (A) Naïve avoidance of electric shocks with different intensities. (B) Single training with increasing voltages leads to increase in both immediate (3 min; left) and 5 h memories (right; n = 10-20). (C) Odor avoidance conditioned with different voltages and tested after 3 min or 5 h correlate with the naïve electroshock avoidance (P = 0.0019 and P = 0.0217 for correlation between naïve avoidance and immediate or 5 h memory, respectively; Pearson's r test). If not stated otherwise, bars and error bars represent means and s.e.m., respectively, in all figures. (B) adapted from Knapek et al., 2010.

To address whether reinforcer intensity affects all distinct memory components equally, I compared the total memory and ARM at 5 hours after training. Cold-shock anesthesia had differential effects on 5-hour memory depending on the intensity of the electric shocks applied during training (Figure 14; $P = 0.001$; significant interaction [shock intensity × anesthesia] with two-way ANOVA; $n \geq 16$). Weak and strong punishment produced similar amounts of ARM ($P > 0.05$; Figure 14), whereas 5-hour memory without cold-shock anesthesia reflected the differences in shock intensities ($P < 0.001$; Figure 14; see also Figure 13B). Altogether, these results suggest that each memory component had a different dependency on shock intensity, raising the question whether their underlying synaptic mechanisms differ.

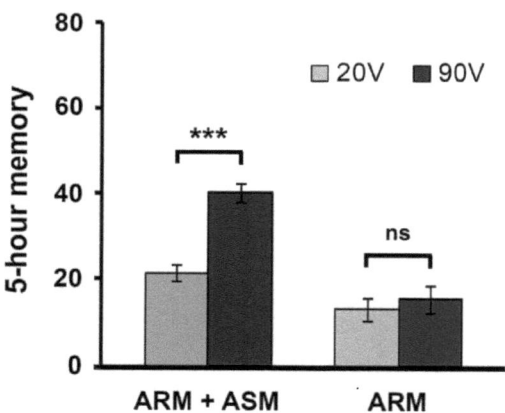

Figure 14: 5 hour memories with different shock intensities. 5 h memory trained with 20 V or 90 V without a cold shock differs significantly ($P < 0.001$; interaction [shock intensity × anaesthesia] with two-way ANOVA, $P = 0.001$, $n \geq 16$; same data as in Figure 13B are presented for the group without cold shock). ARM with weakly reinforced training (20 V) is statistically indistinguishable to that trained at 90 V ($P > 0.05$). The cold shock was applied at three hours after training. Figure modified from Knapek et al., 2010.

3.2 Synapsin is required for short lasting memory

Synapsin is an abundant protein localized to neuropil regions (Figure 15A and E) (Akbergenova and Bykhovskaia, 2007; Angers et al., 2002; Diegelmann et al., 2006a; Greengard et al., 1993; Klagges et al., 1996). We found it highly enriched in presynaptic boutons in the central nervous system (arrowheads in Figure 15B-D and F-H).

Results

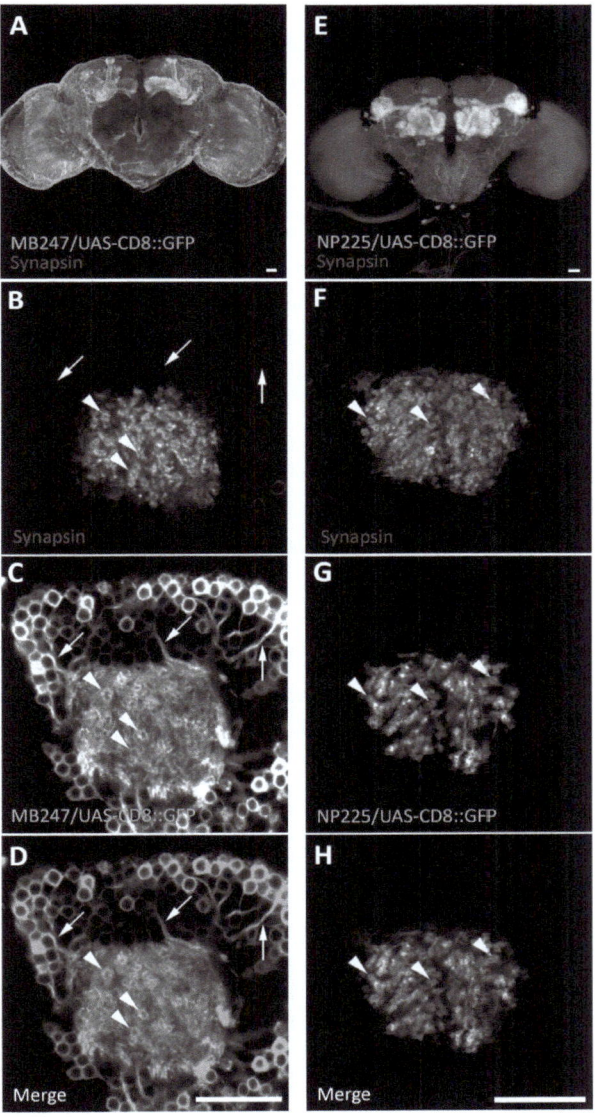

Figure 15: Synapsin is predominantly localized to the presynapses. Magenta: Synapsin; Green: UAS-mCD8-GFP. MB247 (A-D) or NP225 (E-D) drive GAL4 expression predominantly in Kenyon cells of the mushroom body or their presynaptic projection neurons, respectively. Confocal projections of whole brains (A and E) or single slices at the level of the calyx, where projection neurons synapse onto Kenyon cells (B-D and F-H). Synapsin immunopositive large boutons (arrowheads) are surrounded by Kenyon cell dendrites (B-D) and colocalized with the presynaptic terminals of projection neurons (F-H). Neither axons nor cell bodies of Kenyon cells contain Synapsin (arrows). Scale bars: 20μm.

As a prerequisite for attributing a phenotype in memory performance to an impaired learning ability, it is essential to test the responses of the animals to the to-be-associated stimuli. However, the impairment in odor-induced habituation in the *synapsin* mutant (Godenschwege et al., 2004) implies its contribution to the experience-dependent odor response. Therefore I also checked, in addition to the naïve behavior, responses after training-like exposure to either the odors or the shock (sham-training, (Michels et al., 2005; Preat, 1998) and found that responses to several odors after sham-training were significantly different in the *synapsin* mutant (data not shown). Based on this, I chose benzaldehyde and limonene (Figure 16). With this odor combination, neither naïve nor experience-dependent odor responses of the mutant differed from the wild-type control ($P > 0.05$; Figure 16A). In addition, the shock avoidance of wild-type and *synapsin* mutant was indistinguishable, as reported ($P > 0.05$; Figure 16B; Godenschwege et al., 2004).

Given the conditional requirement of Synapsin in vesicle release specifically under high frequency stimulation (Fdez and Hilfiker, 2006; Pieribone et al., 1995; Rosahl et al., 1995; Sun et al., 2006), I hypothesized that Synapsin-dependent vesicle release underlies a specific form of odor memory. Therefore, I examined the immediate and 5-hour memory of the *synapsin* mutant after a single training. Immediate memory of the *synapsin* mutant was, as previously reported, significantly lower than the wild-type control (Figure 16D, $P < 0.01$, $n = 8$) (Godenschwege et al., 2004). This deficit was not detectable when the flies were tested at 5 hours after training (Figure 16C, $P > 0.05$, $n = 8$; $P < 0.01$, interaction [genotype × retention interval] with two-way ANOVA).

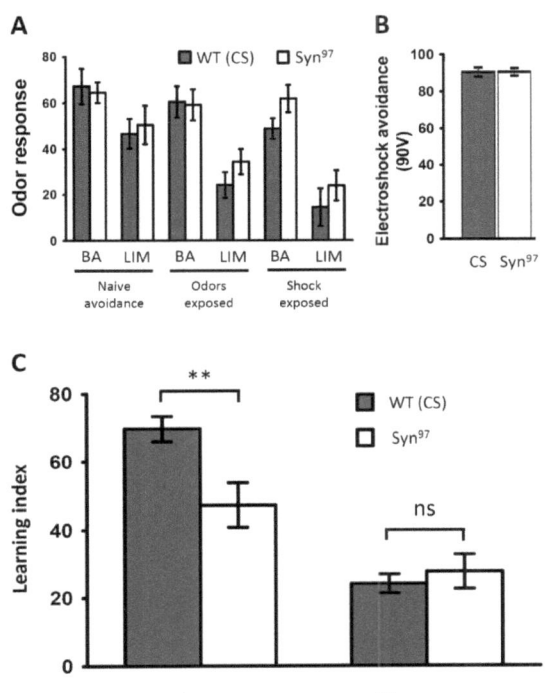

Figure 16: Synapsin is required for immediate, but not for longer lasting memory. (A) Odor responses to the pure chemicals benzaldehyde and limonene. Naïve performance of wild-type flies and the *synapsin* mutant is indistinguishable with this pair of odors. After pre-exposure to either odors or electric shocks, odor responses of the *synapsin* mutant are not significantly different from wild-type CS flies ($P > 0.05$, $n \geq 8$). (B) Shock avoidance (90V) of the *synapsin* mutant is not statistically different from wildtype flies ($P > 0.05$, $n = 12$). (C) The *synapsin* null-mutant shows a significantly decreased immediate memory compared to wild-type flies when trained with these odors ($P < 0.01$, $n = 8$). This deficit is undetectable when the flies are tested 5 hours after training ($P > 0.05$, $n = 8$). Figure modified from Knapek et al., 2010.

Since the conditioned behavior can vary with types and concentration of the respective odors (Akalal et al., 2006; Keene et al., 2004; Tully and Quinn, 1985), I attempted to confirm the selective memory deficit of the *synapsin* mutant with another odor combination (benzaldehyde and 4-methylcyclohexanol) at a lower concentration. To this odor pair, *synapsin* mutants and wild-type flies also did not respond differently (Figure 17A, $P > 0.05$, $n \geq 8$), although they were potentially unable to smell after odor or shock exposure (Figure 17A). Similarly, I found a memory deficit, which was most obvious at short retention intervals of three minutes to up to one hour (Figure 17B, $P < 0.001$ at 3 minutes, $P < 0.01$ at 30 minutes, $P < 0.001$ at 1 hour, $P > 0.05$ at 3 hours, $n \geq 13$). Altogether, these results suggest that the Synapsin-dependent mechanism is selectively required for short-lasting memory.

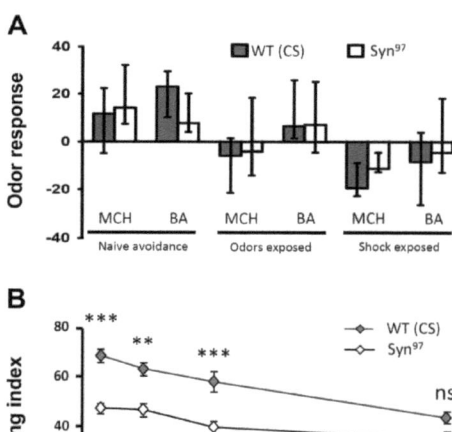

Figure 17: Temporal dynamics of memory decay in wild-type flies and a *synapsin* mutant. (A) Odor response to another pair of diluted odors (4-methylcyclohexanol [1:80] and benzaldehyde [1:1000]), presented in 14mm cups. Naïve performance of wild-type flies and the *synapsin* mutant, as well as the performance after either odor or shock pre-exposure, is indistinguishable with this pair of odors. ($P > 0.05$, $n \geq 8$). Bars and error bars represent median and interquartile range, respectively. (B) Memory retention of the *synapsin* mutant and wild-type flies. Also with this pair of odors, the performance of the mutant in short-lasting memory is significantly reduced (P at least < 0.01 in immediate, 30-minutes and 1-hour memory), whereas 3-hour memories show no statistically significant difference ($P > 0.05$, $n \geq 13$).

3.3 Synapsin is specifically required for anesthesia-sensitive memory

Since short-term olfactory memory mainly consists of a labile memory component (i.e. ASM), I hypothesized that Synapsin is specifically required for ASM. To this end, I dissected the composition of two-hour memory of wild-type flies and the *synapsin*^{97} mutant by administering cold-anesthesia at 1 hour after training (Figure 18). The effect of this anesthetic treatment on 2-hour memory after training with benzaldehyde and limonene was different between these genotypes (Figure 18A, $P < 0.001$; interaction [cold-shock × genotype] with two-way ANOVA, $n \geq 11$). Without cold-anesthesia, the memory score of the *synapsin* mutants was lower than in wild-type flies ($P < 0.001$), whereas ARM of these flies was indistinguishable ($P > 0.05$). These results were reproduced with the above-mentioned diluted second pair of odors benzaldehyde and 4-methylcyclohexanol (Figure 18B).

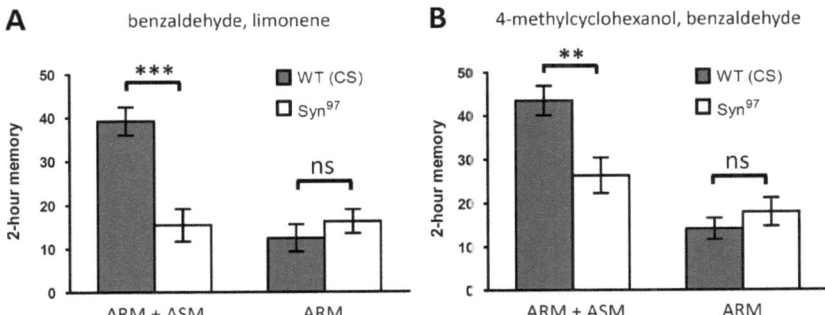

Figure 18: Synapsin contributes to anesthesia-sensitive but not anesthesia-resistant memory. (A) Without cold-anesthesia, memory scores after training with benzaldehyde and limonene are lower in the *synapsin* mutant ($P < 0.001$). When cold-anesthesia was applied at one hour after training, no significant difference in ARM was detected ($P > 0.05$; $n \geq 11$). (B) With diluted benzaldehyde and 4-methylcyclohexanol as odors, this result was reproduced. Total 2-hour memory is significantly impaired in synapsin mutants ($P < 0.01$), whereas ARM is intact ($P > 0.05$, $n = 15$). Thus, the memory component that is missing in the *synapsin* mutants is sensitive to cold-anesthesia. Anesthesia was induced at one hour after training. (A) modified from Knapek et al., 2010.

The effect of the anesthetic treatment on 2-hour memory was different between wild-type and the *synapsin* mutant ($P < 0.01$; interaction [cold-shock × genotype] with two-way ANOVA, $n = 15$). Total memory of *synapsin* mutants was lower ($P < 0.01$), whereas ARM was indistinguishable ($P > 0.05$). Altogether, this suggests that the short-lasting memory impairment in *synapsin* mutants is due to a selective deficit in ASM.

3.4 Synapsin as a potential target of PKA

To genetically address whether cAMP/PKA signaling operates through Synapsin, I asked whether the memory phenotype of the *synapsin* mutant is additive to the *rutabaga* and *dunce* mutants (Byers et al., 1981; Levin et al., 1992). Both genes encode components of the cAMP/PKA cascade and are involved in early memories (Figure 19, see also earlier reports: (Dudai et al., 1976; Tully and Quinn, 1985). If Synapsin acts downstream of the cAMP/PKA cascade, the absence of Synapsin protein should not matter in the *rutabaga or dunce* mutant background; that is, the double mutants should not be impaired beyond the more-severely affected single-mutant. Indeed, short-term memory of the double mutants is not significantly lower than the *rutabaga^{2080}* or *dunce^1* single mutant, respectively (Figure 19A). All single mutants performed significantly worse than the wild-type control ($P < 0.001$, $n \geq 12$). I reproduced these findings in 30-minute memory with the above-mentioned second odor pair (benzaldehyde and 4-methylcyclohexanol). Both double mutants did not perform significantly worse than the respective single mutants (Figure 19B, $P > 0.05$ each), while the three tested single mutants performed significantly worse than the wild-type control ($P < 0.05$ for *synapsin*, $P < 0.001$ for *dunce* and *rutabaga*, $n \geq 9$). These results imply that Synapsin is part of the cAMP cascade and therefore, if they function in the same cells, potentially acts as one of the downstream targets of PKA.

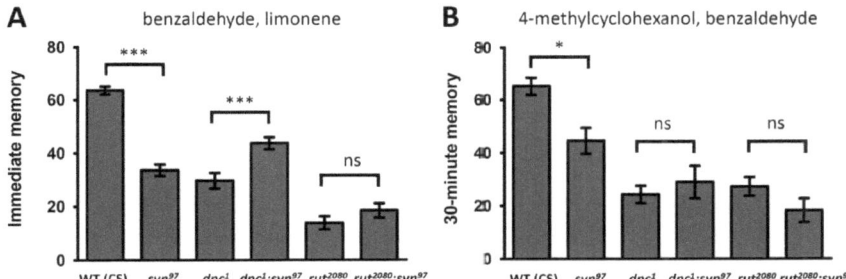

Figure 19: No additive learning defect of mutations in the *synapsin* and *rutabaga* or *dunce* genes. Compared to wild-type flies, *synapsin* mutants show significantly less immediate (A) and 30-minute (B) memory ($P < 0.001$ and $P < 0.05$, respectively). *synapsin/rutabaga* and *synapsin/dunce* double mutants show no additive memory defects compared to the single-mutant for *rutabaga* or *dunce*, respectively ($P > 0.05$, $n \geq 12$ and $n \geq 9$ for immediate (A) and 30-minute (B) memory). Thus, *synapsin* might act as a part of the cAMP/PKA cascade, rather than in a parallel pathway. Please note that *synapsin/dunce* performed significantly better than the single mutants. Benzaldehyde/limonene were used as odors in (A), 4-methylcyclohexanol/benzaldehyde in (B). (A) modified from Knapek et al., 2010.

Synapsin-dependent maintenance of synaptic release is selectively impaired under high-frequency nerve stimulation (Akbergenova and Bykhovskaia, 2007; Gitler et al., 2004; Pieribone et al., 1995; Rosahl et al., 1993; Sun et al., 2006). Phosphorylation of Synapsin after global presynaptic Ca^{2+} increase promotes the recruitment of vesicles from the reserve pool for subsequent release (Bloom et al., 2003; Gaffield and Betz, 2007; Menegon et al., 2006; Pieribone et al., 1995; Sun et al., 2006). This could be a biochemical explanation for the ASM-specific deficit of the *synapsin* mutant. If so, synaptic vesicles that directly at the release sites form the so-called readily releasable pool might be sufficient for ARM. Therefore, I selectively manipulated this complementary vesicle pool by interfering with the presynaptic active zone protein Bruchpilot.

3.5 Bruchpilot is preferentially required for ARM

As Synapsin is specifically required for ASM, I wondered whether other presynaptic proteins might also show impairments in specific memory components. I decided to test the role of Bruchpilot (Brp), which was shown to be necessary for the formation of T-bars at the active zone and to play a critical role in the regulation of neurotransmitter release. Brp is ubiquitously expressed in the *Drosophila* brain (Figure 20A). As a null mutant for *brp* is not viable to adulthood (Kittel et al., 2006), I chose a transgenic RNA interference (RNAi) approach combined with the UAS/GAL4 system to suppress Brp in a targeted group of cells (Wagh et al., 2006). The employed *OK107* GAL4 line drives reporter expression in the majority of MB

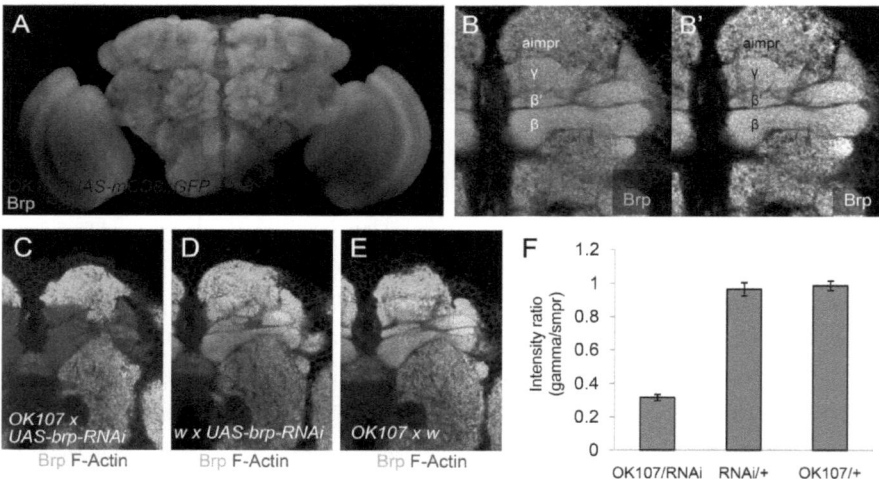

Figure 20: Verification of the knock-down of Brp in the Kenyon cells of the mushroom body. (A) Expression of Brp (green) and *OK107* (blue) in the frontal view of the whole-mount fly brain (projection, dorsal up). Brp is ubiquitously expressed in the brain and localized to the neuropile regions. (B) A magnified confocal slice where the different lobes of the mushroom body (γ, β' and β) and the adjacent neuropile (anterior inferior medial protocerebrum, aimpr) are readily discernible. (C-E) Magenta: F-actin; Green: Brp. (C-E) Brp knock-down by *brp*-RNAi in the mushroom body using *OK107* (C) and genetic controls (D) and (E). Brp and F-actin are visualized in green and magenta, respectively. (F) Intensity ratio of Brp signals in the γ neurons of the Kenyon cells and the adjacent control region (aimpr). After the knock-down, the Brp signal in the mushroom body was significantly reduced to around 30 % ($P < 0.001$, $n = 8$-17). Figure from Knapek et al., 2011.

Kenyon cells (Figure 20A) (Aso et al., 2009; Connolly et al., 1996) that are thought to be the location of associative plasticity (Gerber et al., 2004. Heisenberg, 2003). I verified the knock-down by quantifying Brp protein levels using immunohistochemistry. Brp signal in the mushroom body of knock-down flies was significantly reduced to around 30 % of genetic controls (Figure 20F, $P < 0.001$, $n = 8$-17). Residual Brp signal was expected, because the lobe of the mushroom body also contains many presynaptic terminals of extrinsic neurons that are not affected by *OK107*. Different RNAi constructs that target distinct fragments of the *brp* gene (*brp-RNAi^{C8}* and *brp-RNAi^{B3}*; (Wagh et al., 2006) caused similar knock-down effects (data not shown).

Next, I addressed the function of Brp in the mushroom body for olfactory aversive memory. For this purpose, I chose three different UAS-*brp*-RNAi lines as effectors. Two of them carry a construct targeting different computed open reading frames (CG30336 and CG30337, *brp*-RNAi^{C8} and *brp*-RNAi^{B3}, respectively) within the *brp* transcription unit, the third one bears both of these constructs (UAS-*brp*-RNAi^{B3,C8}). Irrespective of the employed *brp*-RNAi constructs, the *OK107*-GAL4 induced Brp knockdown in the Kenyon cells led to a significantly impaired memory tested at 3 hours after training ($P < 0.01$, $n = 17$-29; Figure 21A). As 3 hour memory consists of both labile and consolidated memory (i.e. ASM and ARM, respectively), we selectively disrupted ASM by giving cold-anesthesia. ARM was also significantly impaired (Figure 21B; $P < 0.01$, $n = 15$-24). The decrease in ARM was similar to that in total 3h memory (Figure 21). This suggests that Brp-knock down leaves ASM intact. Since *brp-RNAi^{C8}* and *brp-RNAi^{B3}*, as well as the double insertion showed the same phenotype in ARM (Figure 21 and Figure 22), the defect is likely due to silencing of Brp.

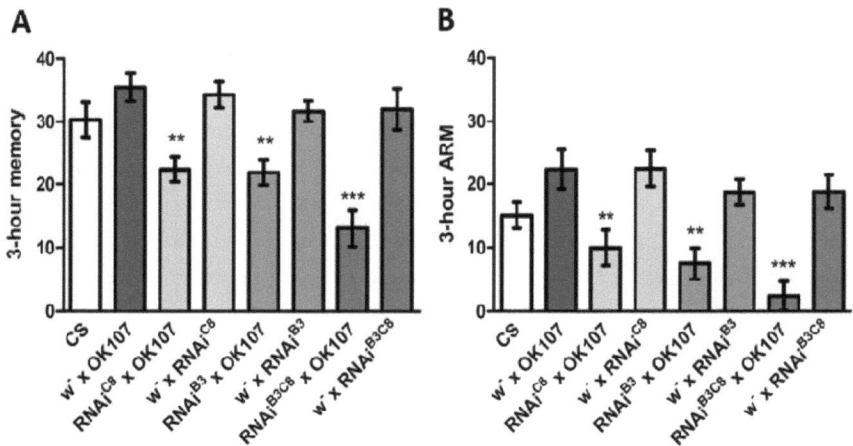

Figure 21: Bruchpilot is specifically required for the consolidated ARM. (A) Brp knock-down in the mushroom body with three different UAS-*brp*-RNAi lines driven by *OK107*-GAL4 leads to a significantly impaired ARM at 3 hours after training ($P < 0.05$; $n = 15$-24). Flies were subjected to cold-anesthesia at two hours after training. (B) Total 3-hour memory (ARM + ASM) in Brp knocked down flies is similarly impaired as ARM alone ($P < 0.01$; $n = 17$-29), suggesting a specific requirement of Brp for ARM. *P*-values refer to the least significant difference of the two respective comparisons (*brp*-RNAi/+ and *OK107*-GAL4/+). Indicated significance level of stars: * ≤ 0.05; ** ≤ 0.01; *** ≤ 0.001. Figure modified from Knapek et al., 2011.

The Brp knock-down in the mushroom body did not lead to a significant reduction of avoidance of electric shock or odors as compared to the corresponding controls (Table 2). Together with the normal ASM (Figure 21 and Figure 22), these results suggest a preferential requirement of Brp for ARM.

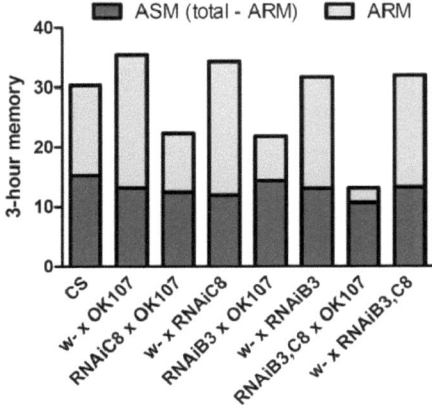

Figure 22: **Specific requirement of Brp for ARM.** Another presentation of the data from **Figure** 21 (A) and (B). The score of ASM (calculated by subtracting ARM from total 3 h memory) is similar in all groups, whereas ARM is specifically impaired in the experimental groups. Modified from Knapek et al., 2011.

3.6 Bruchpilot requirement for ARM is specific to the mushroom body

As most GAL4 driver lines have additional expression outside of the investigated structures (Aso et al., 2009), I tried to confirm the specificity of the RNAi effect to the mushroom bodies with two different strategies: (1) Brp knockdown induced by different GAL4 drivers with overlapping expression in the mushroom body of the Kenyon cells and (2) blocking GAL4 transactivation in the mushroom body using *MB247-GAL80* (Krashes et al., 2007).

In parallel to the *brp* knock-down with *OK107*, I examined the drivers *MB247* with strong expression in the α/β and γ neurons (Zars et al., 2000), and *c739* with strong expression in the α/β neurons (Yang et al., 1995). *brp*-RNAi driven by *OK107* and *c739* led to a significant impairment in total 3-hour memory compared to the two corresponding controls (Figure 23A; $P < 0.001$ and $P < 0.01$, respectively, $n = 16\text{-}55$) whereas flies with *brp*-RNAi driven by MB*247* were not significantly impaired compared to the UAS-*brp*-RNAi control ($P > 0.05$). This might be due to a generally better learning performance of flies containing the MB*247* construct. With all three drivers, flies with *brp* knock-down showed a significant defect in ARM (Figure 23B; $n = 18\text{-}43$, $P < 0.001$, $P < 0.01$, $P < 0.001$ for *OK107*, *MB247* and *c739*, respectively). This result is consistent with the requirement of the α/β neuron output for ARM (Isabel et al., 2004). *brp* knock-down with any of the employed GAL4 lines did not significantly alter the response to electric shocks or odors compared to the corresponding controls ($P > 0.1$; $n = 7$-20 except for the comparison *OK107*-GAL4 x UAS-*brp*-RNAi^{B3C8} vs. w^- x UAS-*brp*-RNAi^{B3C8} in 3-octanol response, $P < 0.05$; see also Table 2).

In the second approach, GAL4 expression of *OK107*-GAL4 and *c739*-GAL4 was suppressed in the mushroom body by *MB247*-GAL80 (Krashes et al., 2007). Therefore, only in the non-overlapping regions of *MB247* and the two respective GAL4 lines (i.e. mostly outside of the mushroom body) the RNAi should be expressed to the full extent. As expected, addition of *MB247-GAL80* significantly improved ARM of the knock-down with *OK107* or *c739* (Figure 23C, $P < 0.01$, $P < 0.05$, for *OK107*-GAL4 and *c739*-GAL4, respectively). ARM with *MB-247-GAL80* was also significantly lower than the control groups without RNAi ($P < 0.05$, $P < 0.01$, *OK107* and *c739*, respectively, $n = 14$-42). The effect of *MB247-GAL80* may therefore be incomplete. Consistently, a reduction of Brp in the mushroom body of *OK107/MB247-GAL80* could still be detected (data not shown). Altogether, these two different experiments suggest that ARM requires a Brp-dependent mechanism at the mushroom body synapses, possibly in the α/β neurons.

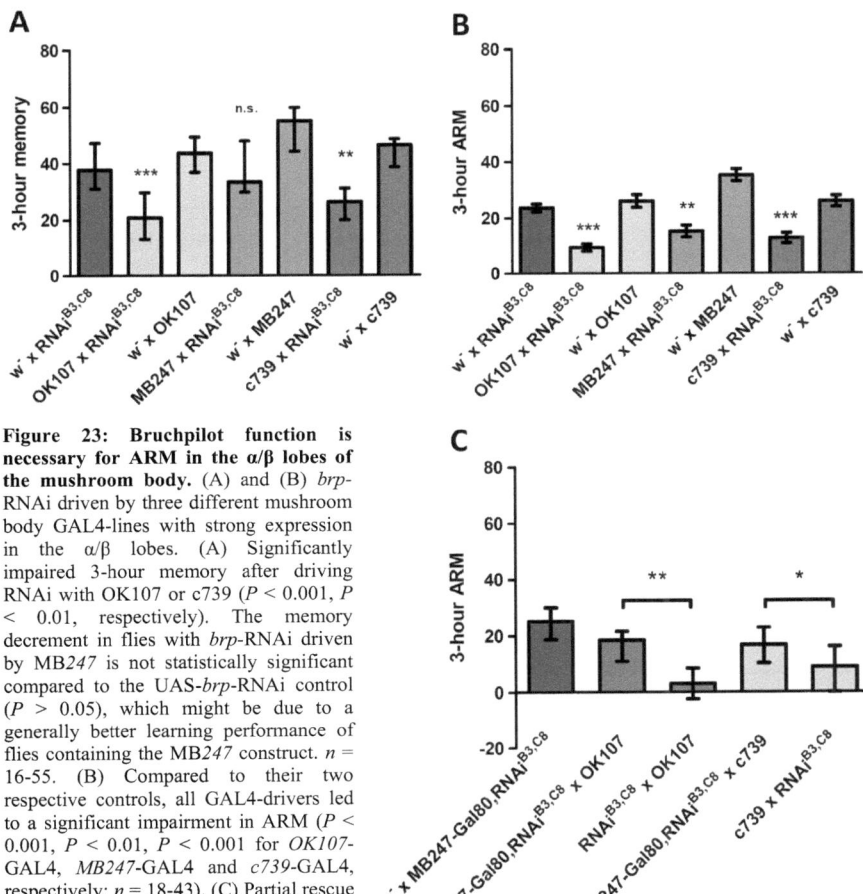

Figure 23: Bruchpilot function is necessary for ARM in the α/β lobes of the mushroom body. (A) and (B) *brp*-RNAi driven by three different mushroom body GAL4-lines with strong expression in the α/β lobes. (A) Significantly impaired 3-hour memory after driving RNAi with OK107 or c739 ($P < 0.001$, $P < 0.01$, respectively). The memory decrement in flies with *brp*-RNAi driven by MB*247* is not statistically significant compared to the UAS-*brp*-RNAi control ($P > 0.05$), which might be due to a generally better learning performance of flies containing the MB*247* construct. n = 16-55. (B) Compared to their two respective controls, all GAL4-drivers led to a significant impairment in ARM ($P < 0.001$, $P < 0.01$, $P < 0.001$ for *OK107*-GAL4, *MB247*-GAL4 and *c739*-GAL4, respectively; n = 18-43). (C) Partial rescue of the ARM defect by blocking GAL4 expression in the mushroom body with GAL80 under control of the *MB247* promoter ($P < 0.01$, $P < 0.05$, for *OK107*-GAL4 and *c739*-GAL4, respectively; n = 14-42). Note that ARM is still significantly lower than the control group without any expression of *brp*-RNAi ($P < 0.05$, $P < 0.01$, *OK107*-GAL4 and *c739*-GAL4, respectively, n = 14-42). As the data of (A and C) are not normally distributed, bars and error bars represent median and interquartile range, respectively. (B) and (C) from Knapek et al., 2011.

Table 2: Naïve response to electric shocks and odors of crosses used in Figure 22 and Figure 23.

Flies expressing the UAS-*brp*-RNAi^{B3C8} construct with different mushroom body GAL4 lines do not show a significantly altered shock or odor response of the tested stimuli ($P > 0.1$; $n = 7\text{-}20$). Only exception was the comparison between *OK107*-GAL4 x UAS-*brp*-RNAi^{B3C8} and w^- x UAS-*brp*-RNAi^{B3C8} in the 3-octanol response (marked with * in the table), $P < 0.05$. As some of the data for shock response were not normally distributed, non-parametric statistics were applied there.

	shock response (median [25%/75% percentile])	3-octanol response (mean [s.e.m.])	4-methylcyclohexanol response (mean [s.e.m.])
OK107 x RNAi^{B3C8}	64.68 [56.22/70.41]	*24.36 [5.70]	31.27 [5.86]
w^- x OK107	70.03 [67.98/77.29]	32.46 [6.60]	25.31 [5.06]
MB247 x RNAi^{B3C8}	69.70 [62.50/73.24]	49.15 [8.00]	58.96 [5.11]
w^- x MB247	66.35 [61.79/71.96]	33.73 [5.45]	40.27 [6.04]
c739 x RNAi^{B3C8}	61.90 [48.15/62.96]	32.33 [5.79]	41.05 [5.78]
w^- x c739	62.07 [53.78/65.99]	29.11 [6.05]	35.46 [6.78]
w^- x RNAi^{B3C8}	66.14 [58.75/80.62]	*36.08 [3.75]	56.62 [3.44]

3.7 Bruchpilot knock-down reduces immediate memory

I showed that at 3 hours after training, Bruchpilot is specifically required for ARM (Figure 21). As immediate memory is completely sensitive to cold-induced anesthesia (and thus ARM is not existing at that early timepoint after training, see Figure 8), Brp should not be required then. To test this hypothesis, I knocked-down Brp in the mushroom body by expressing the brp-RNAi^{B3C8} construct with $OK107$-GAL4 and measured 3-minute memory (Figure 24). Interestingly, immediate memory of these flies was reduced by nearly 50% ($P < 0.001$, $n = 14$). This suggests that immediate memory consists of at least two components, a Brp-dependent component and a Brp-independent one. The Brp-dependent fraction of immediate memory might therefore be consolidated to ARM.

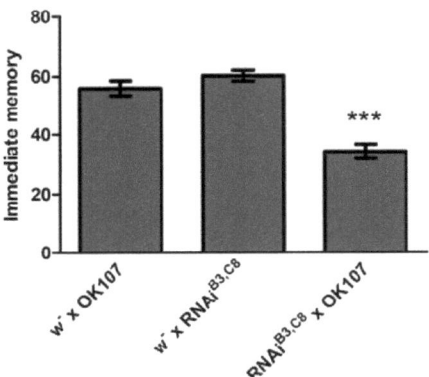

Figure 24: Bruchpilot knock-down affects immediate memory. Expression of brp-RNAi in the mushroom body significantly reduces 3-minute memory compared to controls ($P < 0.001$, $n = 14$).

3.8 Bruchpilot knock-down and *rutabaga* mutant show additive memory impairment

The previous results indicated that Synapsin dependent regulation of the reserve pool is preferentially required for ASM and that Synapsin is part of the *rut*-dependent cAMP pathway (see above). To address whether the Brp-dependent ARM is also regulated by the cAMP pathway, I knocked down Brp in the mushroom body and tested whether the memory impairment of these flies is additive to the one of the *rutabaga* mutant (Figure 25). Unlike the controls, *rutabaga* mutant flies with knocked-down Brp in the mushroom body showed no detectable memory at 3 hours ($P > 0.05$). The memory defect was more severe than the impairment of either *rutabaga* mutation or *brp*-RNAi alone ($P < 0.05$, $n = 34$-57; Figure 25). The avoidance of shock and odors (benzaldehyde and 3-octanol) was not significantly impaired in *rutabaga*; *brp*-RNAi flies (Table 3). Altogether, these results are in line with the hypothesis that ASM and ARM may trigger distinct mechanisms of vesicle release via parallel intracellular signaling pathways.

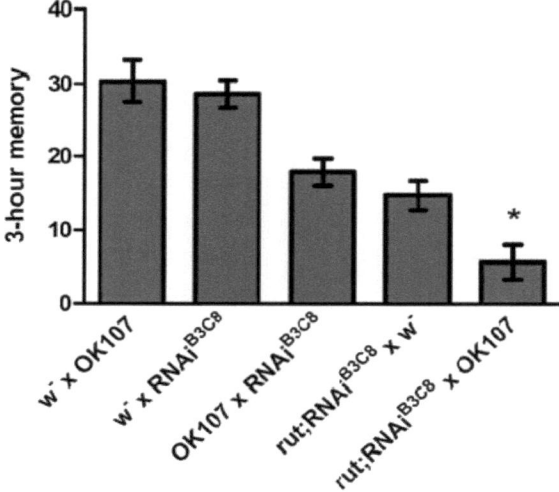

Figure 25: **Additive memory impairment in a *rutabaga* mutant and Bruchpilot knock-down flies.** Flies expressing *brp*-RNAi in the mushroom body in a *rutabaga* mutant background perform significantly worse in 3 h memory than flies with defects in either the Rutabaga or Brp level ($P < 0.05$, $n = 34$-57). Unlike the other groups, 3 h memory of these double mutants was not significantly different from zero ($P > 0.05$). As the response to 4-methylcyclohexanol was impaired in *rut*; *brp*-RNAi flies, benzaldehyde and 3-octanol were used as odors for this experiment. Modified from Knapek et al., 2011.

Table 3: Naïve response to electric shocks and odors of crosses used in Figure 25.

	shock avoidance (mean [s.e.m.])	3-octanol avoidance (mean [s.e.m.])	benzaldehyde avoidance (mean [s.e.m.])
OK107 x RNAi^{B3C8}	41.45 [6.94]	11.51 [4.84]	26.55 [6.99]
rut; RNAi^{B3C8} x OK107	46.55 [6.88]	13.51 [4.80]	15.02 [4.33]
rut; RNAi^{B3C8} x w^-	63.52 [4.14]	-2.88 [6.83]	28.21 [4.55]

Shock and odor responses or flies with impaired Rutabaga and Bruchpilot levels are not significantly reduced compared to the *rutabaga* single mutant or the flies with only low levels of Brp in the mushroom body ($P > 0.05$, $n = 12$).

4 Discussion

4.1 Reinforcer intensity and memory retention

Olfactory memories of fruit flies depend on reinforcer intensity as well as on the number of training cycles (Beck et al., 2000; Diegelmann et al., 2006a; Dudai et al., 1988; Tempel et al., 1983; Tully and Quinn, 1985; Figure 13). This dependency seems to be different in each memory component. For example, ASM at five hours after training is more amenable to electric shock intensity than the consolidated ARM (Figure 14). Therefore, besides the quality of reinforcement (i.e. appetitive or aversive), also the intensity of the reinforcement differentiates olfactory memories. Since flies rarely encounter such life-threatening events as high-voltage electric shocks in their normal life and since formation of longer-lasting memory can be costly (Mery and Kawecki, 2005), it seems reasonable that the strong reinforcement preferentially supports the labile memory ASM. In other words, ARM might be a prioritized memory component, because weak reinforcement was sufficient to induce 5-hour memory (Figure 14).

In *Aplysia* sensorimotor synapses, the released neuromodulator serotonin seems to represent reinforcer intensity. Depending on the repetition and amount of serotonin, different forms of synaptic plasticity, are induced (Emptage and Carew, 1993; Sherff and Carew, 2004). Similarly, the amount of released dopamine might differentiate distinct olfactory memory components in *Drosophila*, where this neurotransmitter was shown to signal negative reinforcement (Gerber et al., 2004; Schroll et al., 2006; Schwaerzel et al., 2003).

4.2 Synapsin mediates the labile memory ASM

Lack of Synapsins in *Drosophila* as well as in vertebrates was found to cause learning defects (Godenschwege et al., 2004; Michels et al., 2005; Silva et al., 1996). My results further refined the role of Synapsin by showing that the memory deficit of *synapsin* is specific to the

ASM (Figure 18). In addition, the result that the mutant defect in immediate memory did not affect the consolidated ARM (Figure 18) supports a model of parallel formation of ARM and ASM (Isabel et al., 2004). ARM is formed without the labile Synapsin-dependent ASM. Such parallel formation of different memory components is also evident at the *Aplysia* sensorimotor synapses mentioned above. With certain stimulation, the long-term facilitation can be induced without short-term facilitation (Emptage and Carew, 1993; Sherff and Carew, 2004).

The molecular mechanism of Synapsin may be important to reveal a synaptic correlate for ASM formation, especially at the level of synaptic vesicle release. In contrast to its ubiquitous neuronal expression, the Synapsin-dependent release of synaptic vesicles was shown to be selective under high-frequency nerve stimulation (Akbergenova and Bykhovskaia, 2007; Gitler et al., 2004; Godenschwege et al., 2004; Pieribone et al., 1995; Rosahl et al., 1993; Sun et al., 2006). Phosphorylation of Synapsin, triggered by global Ca^{2+} increase in presynaptic boutons, promotes vesicle release from the reserve pool, which is located distant from the actual release sites (Bloom et al., 2003; Gaffield and Betz, 2007; Menegon et al., 2006; Pieribone et al., 1995; Sun et al., 2006). This conditional vesicle recruitment could be a biochemical explanation for the selective memory deficit of the *synapsin* mutant. The Synapsin-dependent recruitment of reserve pool vesicles may be necessary for the full level of ASM.

Synapsin appears to be ubiquitously expressed in the *Drosophila* brain (Figure 15A). Identification of the cells where Synapsin functions presynaptically for olfactory memory will help identifying the corresponding output synapses essential for ASM formation (Figure 15B-D, F-H). Because this study employed a null-mutant for *synapsin*, the spatial requirement could not be directly answered. Nevertheless, I can exclude several types of neurons from the candidate sites. All neurons mediating the unconditioned response to the odors and the electric shocks may be ruled out, because the *synapsin* mutants exhibited normal naïve responses to the applied odors and electric shocks (Figure 16A, B and Figure 17A). Thus, there may be three groups of candidate neurons left. Like in the *Aplysia* sensorimotor synapses discussed above, the first possibility is in the reinforcing neurons (Angers et al., 2002). In olfactory learning, dopaminergic neurons signal negative reinforcement of electric shocks (Schroll et al., 2006; Schwaerzel et al., 2003). Another possibility is the output synapses of the Kenyon cells. These synapses undergo associative plasticity underlying short-lasting memory upon the simultaneous presentation of odor and shock (Gerber et al., 2004). Synapsin may regulate vesicle release for this associative plasticity. The remaining potential neuronal pathway are the neurons transmitting the plasticity from the Kenyon cells, such as output neurons from the mushroom body (Tanaka et al., 2008). Synapsin-dependent vesicle release may be specifically required to elicit the conditioned response. Spatially restricted manipulation, such as GAL4-inducible RNAi or tissue-specific rescue experiments, would be required to localize the synapses where Synapsin functions.

4.3 Synapsin as a potential phosphorylation target of PKA

In this thesis, I investigated the epistasis of *synapsin* and components of the cAMP cascade by comparing the memory performance of double-mutants for *synapsin/rutabaga* and *synapsin/dunce* with the respective single mutants (Figure 19). Because the memory deficits of the single mutants were not additive in the corresponding double mutants, Synapsin may be part of the cAMP/PKA signaling. This hypothesis is supported by electrophysiology. At the *Drosophila* neuromuscular junction, high-frequency stimulation was shown to promote vesicle release from the reserve pool through cAMP/PKA activity, as this vesicle recruitment from the reserve pool was specifically affected in the *rutabaga* mutant (Kuromi and Kidokoro, 2000, 2002). This regulation of reserve pool vesicles is similar to the function of Synapsin (Akbergenova and Bykhovskaia, 2007). Furthermore, in *synapsin* as well as in *rutabaga* mutants the preferentially affected olfactory memory component is ASM (Figure 18; (Isabel et al., 2004). Altogether, these results suggest that Synapsin is regulated by the cAMP pathway in *Drosophila*.

Does learning-dependent activation of cAMP signaling induce associative plasticity by phosphorylating Synapsin? PKA-dependent phosphorylation of Synapsin has been reported in vertebrates as well as in *Aplysia* (Bonanomi et al., 2005; Fiumara et al., 2004; Menegon et al., 2006). In rodents, phosphorylation of Synapsin I by PKA was shown to change the distribution and recycling of synaptic vesicles (Bonanomi et al., 2005; Menegon et al., 2006). In *Aplysia*, the single phosphorylation site in the conserved domain of Synapsin was found to be an *in vitro* substrate for PKA. A functional effect of Synapsin phosphorylation by PKA on neurotransmitter release was shown by injecting Synapsin with a mutated phosphorylation site in the land snail *Helix pomatia* (Fiumara et al., 2004).

In *Drosophila*, the Synapsin protein has two putative PKA phosphorylation sites (Diegelmann et al., 2006a). The conserved amino terminal phosphorylation motif however undergoes RNA editing, and therefore phosphorylation by PKA is less efficient (Diegelmann et al., 2006a). Thus, Synapsin might be an indirect target of cAMP/PKA signaling. Further studies using mutated phosphorylation sites may clarify the role of the phosphorylation and PKA signaling for memory formation in *Drosophila*.

4.4 Bruchpilot, a new protein for ARM

In *Drosophila*, associative olfactory memory after single training consists of at least two distinct forms of memory, the labile ASM and the consolidated ARM (Quinn and Dudai, 1976). The molecular mechanisms of ARM are poorly understood compared to the formation of ASM, since the number of mutants affecting ARM is limited so far (Folkers et al., 1993; Folkers et al., 2006; Horiuchi et al., 2008; Tully et al., 1994). Here I demonstrated that RNAi-mediated silencing of the presynaptic active-zone protein Brp (Hofbauer et al., 2009; Kittel et al., 2006; Wagh et al., 2006) in the mushroom body selectively reduced ARM (Figure 21 and Figure 23). Interestingly, the Radish protein, which is required for ARM, is also highly expressed in the mushroom body (Folkers et al., 1993). Thus, Brp and Radish might interact at the presynaptic terminals to regulate neurotransmission underlying ARM.

Several memory mutants have been shown to have a selective phenotype in either ARM or ASM. For instance, ASM is specifically affected in mutants with an impaired cAMP/PKA signaling pathway (e.g. *rutabaga*, encoding the type I adenylate cyclase; Dudai et al., 1988; Folkers et al., 1993; Isabel et al., 2004; Schwaerzel et al., 2007). Here I showed that Synapsin, a presynaptic vesicle protein essential for the regulation of reserve pool vesicles, is mainly required for ASM. The preferential ARM impairment of Brp knocked-down flies (Figure 21 and Figure 23) is complementary to the function of Synapsin for ASM. Moreover, *brp* knock-down in *rutabaga* mutant flies leads to an additive memory deficit (Figure 25), whereas the lack of Synapsin causes no enhancement of the memory phenotype in the *rutabaga* mutant. Together, this implies a dissociation of memory components via different signaling pathways and synaptic proteins. However, Brp plays also a significant role in short-term memory (Figure 24), which can be completely disrupted by cold-induced anesthesia (Figure 8). Therefore, a signaling pathway for ASM might regulate Brp in early memory. Such regulation could depend on PKA, as it has been shown recently that PKA can inhibit the formation of ARM (Horiuchi et al., 2008).

4.5 Dissociation of ARM and ASM

ASM dominates early memory and decays quickly, whereas ARM is not detectable immediately after training and gradually develops afterwards (Quinn and Dudai, 1976). Although Radish and Brp are required for ARM, flies lacking either of the proteins are also impaired in their immediate memory (Figure 24; Folkers et al., 1993). Thus, formation of ARM could be either *de novo*, i.e. completely in parallel to ASM, or a fraction of the early labile memory might act as a "precursor" which becomes gradually consolidated. This fraction could be the Rad- and Brp-dependent part of immediate memory.

It has been shown that flies are able to not only form a memory for odor qualities, but also for different odor intensities (Masek and Heisenberg, 2008; Yarali et al., 2009). However, this odor intensity memory (OIM) seems to be not as strong and long-lasting than the odor quality memory (Masek and Heisenberg, 2008). Therefore, the odor quality cues may be more salient than the odor intensity cues for the flies. Although it might well be that the intensity memory is not coexpressed in the presence of the potentially more salient odor quality memory (Tang and Guo, 2001), one could nevertheless assume that odor intensity and odor quality memory together compose the immediate olfactory memory in flies. Interestingly, OIM does not require the *rutabaga* and *dunce* genes, and therefore is supposed to be independent of the cAMP/PKA pathway (Masek and Heisenberg, 2008). I have shown in this work that regulation of Brp, which is preferentially required for ARM at three hours after training, might also be independent of the cAMP signaling (Figure 25). Moreover, OIM decays within about 3 hours

Discussion

after training (Masek and Heisenberg, 2008). In contrast, ARM does not exist immediately after training and just forms gradually. Given this complementary temporal dynamics of OIM decay and ARM formation together with their common putative independency of the cAMP/PKA signaling pathway, one could speculate that OIM might be a "precursor" of the Brp dependent ARM. However, so far it has not been tested whether OIM, similar as ARM, requires Brp.

If ARM and ASM were formed at the same synapses of Kenyon cells (e.g. α/β neurons), how can the two synaptic proteins Brp and Synapsin dissociate these different forms of memory? As Brp and Synapsin are required for different processes of neurotransmission, these two forms of memory might be differentiated by vesicle release. Release of synaptic vesicles was shown to be synapsin-dependent under sustained high-frequency nerve stimulation (Akbergenova and Bykhovskaia, 2007; Gitler et al., 2004; Pieribone et al., 1995; Rosahl et al., 1993; Sun et al., 2006). In contrast, the requirement of Brp is most pronounced in immediate vesicle release (Kittel et al., 2006). In line with this selective requirement, in *brp* mutants the clustering of Ca^{2+} channels to the active zones was impaired and less synaptic vesicles of the readily-releasable pool are accumulated at the release sites (Kittel et al., 2006; Wagh et al., 2006). Thus, the two different modes of neurotransmission (high- vs. low frequency dependent) might differentiate ASM and ARM even if they would coexist within the same synapses (see Figure 8).

I found that for ARM, Brp is necessary in the α/β lobe neurons of the mushroom body (Figure 23). This is consistent with earlier experiments, where reversible blocking of synaptic output from the alpha/beta lobes impaired ARM (Isabel et al., 2004). These lobes are formed through bifurcation of a set of Kenyon cells. As blocking synaptic output from extrinsic neurons innervating a distinct part of the beta lobes results in a specific impairment of ASM (Aso et al., submitted), ASM and ARM could also be spatially separated on a subcellular level within these neurons. This hypothesis would be supported by delayed Ca^{2+} signals after training specifically in neurons innervating preferentially the vertical lobes (Yu et al., 2005). Altogether, the distinction of ARM and ASM may underlie a combination of both molecular and spatial separation.

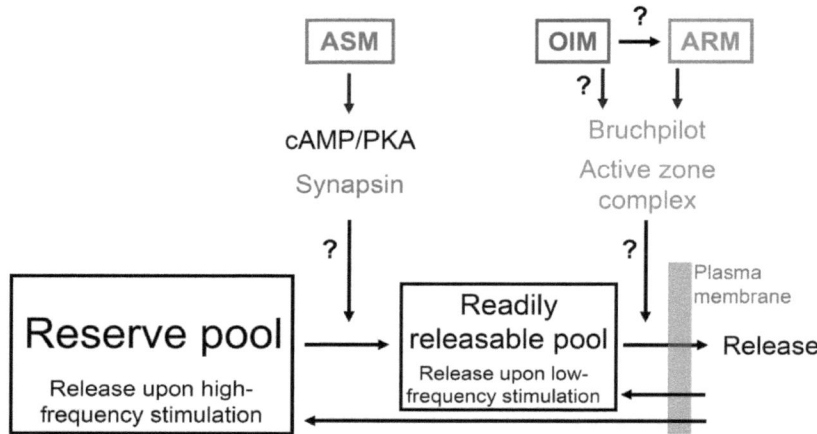

Figure 26: Model of ARM / ASM dissociation. At the presynapse, two separated pools of vesicles exist: a reserve pool situated distantly from the plasma membrane and a readily-releasable pool located directly at the active zones (Kidokoro et al., 2004). Function of the reserve pool associated vesicle protein Synapsin is required for neurotransmitter release selectively under sustained high-frequency nerve stimulation (Akbergenova and Bykhovskaia, 2007; Gitler et al., 2004; Pieribone et al., 1995; Rosahl et al., 1993; Sun et al., 2006). As a *synapsin* mutant is preferentially impaired in ASM, high-frequency evoked release from the reserve pool may specifically contribute to this memory component. In contrast, Brp is required for electron dense structures (T-bars) at the active zone. In *brp* mutants, the clustering of Ca^{2+} channels to the active zones is impaired and less synaptic vesicles of the readily-releasable pool are accumulated at the release sites (Kittel et al., 2006; Wagh et al., 2006). In line with that, requirement of Brp is most pronounced in immediate vesicle release (Kittel et al., 2006). Although Brp is preferentially required for ARM at later timepoints after training, it also plays an important role in immediate memory, which consists entirely of ASM. Because odor-intensity memory (OIM) is also independent of the cAMP/PKA signaling pathway and because its rate of decay is similar to the rate of ARM formation (Masek and Heisenberg, 2008), OIM might be the "precursor" for ARM. Thus, ASM and OIM/ARM might be differentiated by two different modes of neurotransmission, a high-frequency dependent one from the reserve pool and a low-frequency dependent one from the readily-releasable pool.

5 References

Abrams, T.W., and Kandel, E.R. (1988). Is contiguity detection in classical conditioning a system or a cellular property? Learning in Aplysia suggests a possible molecular site. Trends Neurosci *11*, 128-135.

Adams, M.D., Celniker, S.E., Holt, R.A., Evans, C.A., Gocayne, J.D., Amanatides, P.G., Scherer, S.E., Li, P.W., Hoskins, R.A., Galle, R.F., et al. (2000). The genome sequence of Drosophila melanogaster. Science *287*, 2185-2195.

Akalal, D.B., Wilson, C.F., Zong, L., Tanaka, N.K., Ito, K., and Davis, R.L. (2006). Roles for Drosophila mushroom body neurons in olfactory learning and memory. Learn Mem *13*, 659-668.

Akbergenova, Y., and Bykhovskaia, M. (2007). Synapsin maintains the reserve vesicle pool and spatial segregation of the recycling pool in Drosophila presynaptic boutons. Brain Res *1178*, 52-64.

Angers, A., Fioravante, D., Chin, J., Cleary, L.J., Bean, A.J., and Byrne, J.H. (2002). Serotonin stimulates phosphorylation of Aplysia synapsin and alters its subcellular distribution in sensory neurons. J Neurosci *22*, 5412-5422.

Anholt, R.R. (1994). Signal integration in the nervous system: adenylate cyclases as molecular coincidence detectors. Trends Neurosci *17*, 37-41.

Ashburner, M. (1989). Drosophila: A Laboratory Handbook (New York: Cold Spring Harbor Laboratory).

Aso, Y., Grubel, K., Busch, S., Friedrich, A.B., Siwanowicz, I., and Tanimoto, H. (2009). The mushroom body of adult Drosophila characterized by GAL4 drivers. J Neurogenet *23*, 156-172.

Aso Y, Bräcker L, Ito K, Kitamoto T & Tanimoto H. (2009). Identification of specific dopaminergic neurons for aversive memory formation. Curr Biol submitted.

Azevedo, F.A., Carvalho, L.R., Grinberg, L.T., Farfel, J.M., Ferretti, R.E., Leite, R.E., Jacob Filho, W., Lent, R., and Herculano-Houzel, S. (2009). Equal numbers of neuronal and nonneuronal cells make the human brain an isometrically scaled-up primate brain. J Comp Neurol *513*, 532-541.

Balfanz, S., Strunker, T., Frings, S., and Baumann, A. (2005). A family of octopamine [corrected] receptors that specifically induce cyclic AMP production or Ca2+ release in Drosophila melanogaster. J Neurochem *93*, 440-451.

Bausenwein, B., Wolf, R., and Heisenberg, M. (1986). Genetic dissection of optomotor behavior in Drosophila melanogaster. Studies on wild-type and the mutant optomotor-blindH31. J Neurogenet *3*, 87-109.

Beck, C.D., Schroeder, B., and Davis, R.L. (2000). Learning performance of normal and mutant Drosophila after repeated conditioning trials with discrete stimuli. J Neurosci 20, 2944-2953.

Benzer, S. (1967). Behavioral mutants of Drosophila isolated by countercurrent distribution. Proc Natl Acad Sci U S A 58, 1112-1119.

Bier, E. (2005). Drosophila, the golden bug, emerges as a tool for human genetics. Nat Rev Genet 6, 9-23.

Birmingham, A., Anderson, E.M., Reynolds, A., Ilsley-Tyree, D., Leake, D., Fedorov, Y., Baskerville, S., Maksimova, E., Robinson, K., Karpilow, J., et al. (2006). 3' UTR seed matches, but not overall identity, are associated with RNAi off-targets. Nat Methods 3, 199-204.

Bloom, O., Evergren, E., Tomilin, N., Kjaerulff, O., Low, P., Brodin, L., Pieribone, V.A., Greengard, P., and Shupliakov, O. (2003). Colocalization of synapsin and actin during synaptic vesicle recycling. J Cell Biol 161, 737-747.

Blum, A.L., Li, W., Cressy, M., and Dubnau, J. (2009). Short- and long-term memory in Drosophila require cAMP signaling in distinct neuron types. Curr Biol 19, 1341-1350.

Bonanomi, D., Menegon, A., Miccio, A., Ferrari, G., Corradi, A., Kao, H.T., Benfenati, F., and Valtorta, F. (2005). Phosphorylation of synapsin I by cAMP-dependent protein kinase controls synaptic vesicle dynamics in developing neurons. J Neurosci 25, 7299-7308.

Boynton, S., and Tully, T. (1992). latheo, a new gene involved in associative learning and memory in Drosophila melanogaster, identified from P element mutagenesis. Genetics 131, 655-672.

Brand, A.H., and Perrimon, N. (1993). Targeted gene expression as a means of altering cell fates and generating dominant phenotypes. Development 118, 401-415.

Busch, S., Selcho, M., Ito, K., and Tanimoto, H. (2009). A map of octopaminergic neurons in the Drosophila brain. J Comp Neurol 513, 643-667.

Byers, D., Davis, R.L., and Kiger, J.A., Jr. (1981). Defect in cyclic AMP phosphodiesterase due to the dunce mutation of learning in Drosophila melanogaster. Nature 289, 79-81.

Carew, T.J., Pinsker, H.M., and Kandel, E.R. (1972). Long-Term Habituation of a Defensive Withdrawal Reflex in Aplysia. Science 175, 451-454.

Cerutti, H. (2003). RNA interference: traveling in the cell and gaining functions? Trends Genet 19, 39-46.

Cheng, Y., Endo, K., Wu, K., Rodan, A.R., Heberlein, U., and Davis, R.L. (2001). Drosophila fasciclinII is required for the formation of odor memories and for normal sensitivity to alcohol. Cell 105, 757-768.

References

Chi, P., Greengard, P., and Ryan, T.A. (2001). Synapsin dispersion and reclustering during synaptic activity. Nat Neurosci *4*, 1187-1193.

Chi, P., Greengard, P., and Ryan, T.A. (2003). Synaptic vesicle mobilization is regulated by distinct synapsin I phosphorylation pathways at different frequencies. Neuron *38*, 69-78.

Colomb, J., Kaiser, L., Chabaud, M.A., and Preat, T. (2009). Parametric and genetic analysis of Drosophila appetitive long-term memory and sugar motivation. Genes Brain Behav *8*, 407-415.

Connolly, J.B., Roberts, I.J., Armstrong, J.D., Kaiser, K., Forte, M., Tully, T., and O'Kane, C.J. (1996). Associative learning disrupted by impaired Gs signaling in Drosophila mushroom bodies. Science *274*, 2104-2107.

Consortium, I.H.G.S. (2004). Finishing the euchromatic sequence of the human genome. Nature *431*, 931-945.

Couto, A., Alenius, M., and Dickson, B.J. (2005). Molecular, anatomical, and functional organization of the Drosophila olfactory system. Curr Biol *15*, 1535-1547.

Crittenden, J.R., Skoulakis, E.M., Han, K.A., Kalderon, D., and Davis, R.L. (1998). Tripartite mushroom body architecture revealed by antigenic markers. Learn Mem *5*, 38-51.

Davis, R.L. (2005). Olfactory memory formation in Drosophila: from molecular to systems neuroscience. Annu Rev Neurosci *28*, 275-302.

Davis, R.L., Cherry, J., Dauwalder, B., Han, P.L., and Skoulakis, E. (1995). The cyclic AMP system and Drosophila learning. Mol Cell Biochem *149-150*, 271-278.

de Belle, J.S., and Heisenberg, M. (1994). Associative odor learning in Drosophila abolished by chemical ablation of mushroom bodies. Science *263*, 692-695.

Diegelmann, S., Nieratschker, V., Werner, U., Hoppe, J., Zars, T., and Buchner, E. (2006a). The conserved protein kinase-A target motif in synapsin of Drosophila is effectively modified by pre-mRNA editing. BMC Neurosci *7*, 76.

Diegelmann, S., Zars, M., and Zars, T. (2006b). Genetic dissociation of acquisition and memory strength in the heat-box spatial learning paradigm in *Drosophila*. Learn Mem *13*, 72-83.

Dietzl, G., Chen, D., Schnorrer, F., Su, K.C., Barinova, Y., Fellner, M., Gasser, B., Kinsey, K., Oppel, S., Scheiblauer, S., *et al.* (2007). A genome-wide transgenic RNAi library for conditional gene inactivation in Drosophila. Nature *448*, 151-156.

Drain, P., Folkers, E., and Quinn, W.G. (1991). cAMP-dependent protein kinase and the disruption of learning in transgenic flies. Neuron *6*, 71-82.

Dubnau, J., Grady, L., Kitamoto, T., and Tully, T. (2001). Disruption of neurotransmission in Drosophila mushroom body blocks retrieval but not acquisition of memory. Nature *411*, 476-480.

Dubnau, J., and Tully, T. (1998). Gene discovery in Drosophila: new insights for learning and memory. Annu Rev Neurosci *21*, 407-444.

Dudai, Y., Corfas, G., and Hazvi, S. (1988). What is the possible contribution of Ca2+-stimulated adenylate cyclase to acquisition, consolidation and retention of an associative olfactory memory in Drosophila. J Comp Physiol A *162*, 101-109.

Dudai, Y., Jan, Y.N., Byers, D., Quinn, W.G., and Benzer, S. (1976). dunce, a mutant of Drosophila deficient in learning. Proc Natl Acad Sci U S A *73*, 1684-1688.

Duffy, J.B. (2002). GAL4 system in Drosophila: a fly geneticist's Swiss army knife. Genesis *34*, 1-15.

Dujardin, F. (1850). Mémoire sur le système nerveux des insectes. Ann. Sci. Nat. Zool. *14*, 195-206.

Dura, J.M., Preat, T., and Tully, T. (1993). Identification of linotte, a new gene affecting learning and memory in Drosophila melanogaster. J Neurogenet *9*, 1-14.

Ebbinghaus, H. (1885). Über das Gedächtnis. Untersuchungen zur experimentellen Psychologie. (Leipzig: Duncker & Humber).

Echeverri, C.J., and Perrimon, N. (2006). High-throughput RNAi screening in cultured cells: a user's guide. Nat Rev Genet *7*, 373-384.

Emptage, N.J., and Carew, T.J. (1993). Long-term synaptic facilitation in the absence of short-term facilitation in Aplysia neurons. Science *262*, 253-256.

Enerly, E., Larsson, J., and Lambertsson, A. (2003). Silencing the Drosophila ribosomal protein L14 gene using targeted RNA interference causes distinct somatic anomalies. Gene *320*, 41-48.

Farooqui, T., Robinson, K., Vaessin, H., and Smith, B.H. (2003). Modulation of early olfactory processing by an octopaminergic reinforcement pathway in the honeybee. J Neurosci *23*, 5370-5380.

Fdez, E., and Hilfiker, S. (2006). Vesicle pools and synapsins: new insights into old enigmas. Brain Cell Biol *35*, 107-115.

Feany, M.B., and Quinn, W.G. (1995). A neuropeptide gene defined by the Drosophila memory mutant amnesiac. Science *268*, 869-873.

Fiala, A. (2007). Olfaction and olfactory learning in Drosophila: recent progress. Curr Opin Neurobiol *17*, 720-726.

Fiala, A., Spall, T., Diegelmann, S., Eisermann, B., Sachse, S., Devaud, J.M., Buchner, E., and Galizia, C.G. (2002). Genetically expressed cameleon in Drosophila melanogaster is used to visualize olfactory information in projection neurons. Curr Biol *12*, 1877-1884.

Fishilevich, E., and Vosshall, L.B. (2005). Genetic and functional subdivision of the Drosophila antennal lobe. Curr Biol *15*, 1548-1553.

Fiumara, F., Giovedi, S., Menegon, A., Milanese, C., Merlo, D., Montarolo, P.G., Valtorta, F., Benfenati, F., and Ghirardi, M. (2004). Phosphorylation by cAMP-dependent protein kinase is essential for synapsin-induced enhancement of neurotransmitter release in invertebrate neurons. J Cell Sci *117*, 5145-5154.

Folkers, E., Drain, P., and Quinn, W.G. (1993). Radish, a Drosophila mutant deficient in consolidated memory. Proc Natl Acad Sci U S A *90*, 8123-8127.

Folkers, E., Waddell, S., and Quinn, W.G. (2006). The Drosophila radish gene encodes a protein required for anesthesia-resistant memory. Proc Natl Acad Sci U S A *103*, 17496-17500.

Fouquet, W., Owald, D., Wichmann, C., Mertel, S., Depner, H., Dyba, M., Hallermann, S., Kittel, R.J., Eimer, S., and Sigrist, S.J. (2009). Maturation of active zone assembly by Drosophila Bruchpilot. J Cell Biol *186*, 129-145.

Frost, W.N., Castellucci, V.F., Hawkins, R.D., and Kandel, E.R. (1985). Monosynaptic connections made by the sensory neurons of the gill- and siphon-withdrawal reflex in Aplysia participate in the storage of long-term memory for sensitization. Proc Natl Acad Sci U S A *82*, 8266-8269.

Fujishiro, N., Kijima, H., and Miyakawa, Y. (1990). Isolation and characterization of feeding behavior mutants in Drosophila melanogaster. Behav Genet *20*, 437-451.

Gaffield, M.A., and Betz, W.J. (2007). Synaptic vesicle mobility in mouse motor nerve terminals with and without synapsin. J Neurosci *27*, 13691-13700.

Gerber, B., Tanimoto, H., and Heisenberg, M. (2004). An engram found? Evaluating the evidence from fruit flies. Curr Opin Neurobiol *14*, 737-744.

Giordano, E., Rendina, R., Peluso, I., and Furia, M. (2002). RNAi triggered by symmetrically transcribed transgenes in Drosophila melanogaster. Genetics *160*, 637-648.

Gitler, D., Cheng, Q., Greengard, P., and Augustine, G.J. (2008). Synapsin IIa controls the reserve pool of glutamatergic synaptic vesicles. J Neurosci *28*, 10835-10843.

Gitler, D., Takagishi, Y., Feng, J., Ren, Y., Rodriguiz, R.M., Wetsel, W.C., Greengard, P., and Augustine, G.J. (2004). Different presynaptic roles of synapsins at excitatory and inhibitory synapses. J Neurosci *24*, 11368-11380.

Giurfa, M. (2006). Associative learning: the instructive function of biogenic amines. Curr Biol *16*, R892-895.

Godenschwege, T.A., Reisch, D., Diegelmann, S., Eberle, K., Funk, N., Heisenberg, M., Hoppe, V., Hoppe, J., Klagges, B.R., Martin, J.R., et al. (2004). Flies lacking all synapsins are unexpectedly healthy but are impaired in complex behaviour. Eur J Neurosci *20*, 611-622.

Goodwin, S.F., Del Vecchio, M., Velinzon, K., Hogel, C., Russell, S.R., Tully, T., and Kaiser, K. (1997). Defective learning in mutants of the Drosophila gene for a regulatory subunit of cAMP-dependent protein kinase. J Neurosci *17*, 8817-8827.

Goodwin, S.F., Taylor, B.J., Villella, A., Foss, M., Ryner, L.C., Baker, B.S., and Hall, J.C. (2000). Aberrant splicing and altered spatial expression patterns in fruitless mutants of Drosophila melanogaster. Genetics *154*, 725-745.

Greengard, P., Valtorta, F., Czernik, A.J., and Benfenati, F. (1993). Synaptic vesicle phosphoproteins and regulation of synaptic function. Science *259*, 780-785.

Greenspan, R.J., and Ferveur, J.F. (2000). Courtship in Drosophila. Annu Rev Genet *34*, 205-232.

Grotewiel, M.S., Beck, C.D., Wu, K.H., Zhu, X.R., and Davis, R.L. (1998). Integrin-mediated short-term memory in Drosophila. Nature *391*, 455-460.

Hallem, E.A., and Carlson, J.R. (2004). The odor coding system of Drosophila. Trends Genet *20*, 453-459.

Hammer, M. (1993). An identified neuron mediates the unconditioned stimulus in associative olfactory learning in honeybees. Nature *366*, 59-63.

Hammer, M., and Menzel, R. (1998). Multiple sites of associative odor learning as revealed by local brain microinjections of octopamine in honeybees. Learn Mem *5*, 146-156.

Han, D.D., Stein, D., and Stevens, L.M. (2000). Investigating the function of follicular subpopulations during Drosophila oogenesis through hormone-dependent enhancer-targeted cell ablation. Development *127*, 573-583.

Han, K.A., Millar, N.S., and Davis, R.L. (1998). A novel octopamine receptor with preferential expression in Drosophila mushroom bodies. J Neurosci *18*, 3650-3658.

Han, K.A., Millar, N.S., Grotewiel, M.S., and Davis, R.L. (1996). DAMB, a novel dopamine receptor expressed specifically in Drosophila mushroom bodies. Neuron *16*, 1127-1135.

Han, P.L., Levin, L.R., Reed, R.R., and Davis, R.L. (1992). Preferential expression of the Drosophila rutabaga gene in mushroom bodies, neural centers for learning in insects. Neuron *9*, 619-627.

Hearn, M.G., Ren, Y., McBride, E.W., Reveillaud, I., Beinborn, M., and Kopin, A.S. (2002). A Drosophila dopamine 2-like receptor: Molecular characterization and identification of multiple alternatively spliced variants. Proc Natl Acad Sci U S A *99*, 14554-14559.

Heisenberg, M. (2003). Mushroom body memoir: from maps to models. Nat Rev Neurosci *4*, 266-275.

Heisenberg, M., Borst, A., Wagner, S., and Byers, D. (1985). Drosophila mushroom body mutants are deficient in olfactory learning. J Neurogenet 2, 1-30.

Hildebrand, J.G., and Shepherd, G.M. (1997). Mechanisms of olfactory discrimination: converging evidence for common principles across phyla. Annu Rev Neurosci 20, 595-631.

Hilfiker, S., Benfenati, F., Doussau, F., Nairn, A.C., Czernik, A.J., Augustine, G.J., and Greengard, P. (2005). Structural domains involved in the regulation of transmitter release by synapsins. J Neurosci 25, 2658-2669.

Hilfiker, S., Pieribone, V.A., Czernik, A.J., Kao, H.T., Augustine, G.J., and Greengard, P. (1999). Synapsins as regulators of neurotransmitter release. Philos Trans R Soc Lond B Biol Sci 354, 269-279.

Hofbauer, A., Ebel, T., Waltenspiel, B., Oswald, P., Chen, Y.C., Halder, P., Biskup, S., Lewandrowski, U., Winkler, C., Sickmann, A., et al. (2009). The Wuerzburg hybridoma library against Drosophila brain. J Neurogenet 23, 78-91.

Honjo, K., and Furukubo-Tokunaga, K. (2009). Distinctive neuronal networks and biochemical pathways for appetitive and aversive memory in Drosophila larvae. J Neurosci 29, 852-862.

Horiuchi, J., Yamazaki, D., Naganos, S., Aigaki, T., and Saitoe, M. (2008). Protein kinase A inhibits a consolidated form of memory in Drosophila. Proc Natl Acad Sci U S A 105, 20976-20981.

Hosaka, M., and Sudhof, T.C. (1999). Homo- and heterodimerization of synapsins. J Biol Chem 274, 16747-16753.

Isabel, G., Pascual, A., and Preat, T. (2004). Exclusive consolidated memory phases in Drosophila. Science 304, 1024-1027.

Ito, K., Awano, W., Suzuki, K., Hiromi, Y., and Yamamoto, D. (1997). The Drosophila mushroom body is a quadruple structure of clonal units each of which contains a virtually identical set of neurones and glial cells. Development 124, 761-771.

Ito, K., and Hotta, Y. (1992). Proliferation pattern of postembryonic neuroblasts in the brain of Drosophila melanogaster. Dev Biol 149, 134-148.

Ito, K., Okada, R., Tanaka, N.K., and Awasaki, T. (2003). Cautionary observations on preparing and interpreting brain images using molecular biology-based staining techniques. Microsc Res Tech 62, 170-186.

Ito, K., Suzuki, K., Estes, P., Ramaswami, M., Yamamoto, D., and Strausfeld, N.J. (1998). The organization of extrinsic neurons and their implications in the functional roles of the mushroom bodies in Drosophila melanogaster Meigen. Learn Mem 5, 52-77.

References

Jackson, A.L., Burchard, J., Schelter, J., Chau, B.N., Cleary, M., Lim, L., and Linsley, P.S. (2006). Widespread siRNA "off-target" transcript silencing mediated by seed region sequence complementarity. RNA *12*, 1179-1187.

Jefferis, G.S., Marin, E.C., Stocker, R.F., and Luo, L. (2001). Target neuron prespecification in the olfactory map of Drosophila. Nature *414*, 204-208.

Jefferis, G.S., Marin, E.C., Watts, R.J., and Luo, L. (2002). Development of neuronal connectivity in Drosophila antennal lobes and mushroom bodies. Curr Opin Neurobiol *12*, 80-86.

Jefferis, G.S., Potter, C.J., Chan, A.M., Marin, E.C., Rohlfing, T., Maurer, C.R., Jr., and Luo, L. (2007). Comprehensive maps of Drosophila higher olfactory centers: spatially segregated fruit and pheromone representation. Cell *128*, 1187-1203.

Johard, H.A., Enell, L.E., Gustafsson, E., Trifilieff, P., Veenstra, J.A., and Nassel, D.R. (2008). Intrinsic neurons of Drosophila mushroom bodies express short neuropeptide F: relations to extrinsic neurons expressing different neurotransmitters. J Comp Neurol *507*, 1479-1496.

Kalidas, S., and Smith, D.P. (2002). Novel genomic cDNA hybrids produce effective RNA interference in adult Drosophila. Neuron *33*, 177-184.

Kao, H.T., Porton, B., Hilfiker, S., Stefani, G., Pieribone, V.A., DeSalle, R., and Greengard, P. (1999). Molecular evolution of the synapsin gene family. J Exp Zool *285*, 360-377.

Keene, A.C., Krashes, M.J., Leung, B., Bernard, J.A., and Waddell, S. (2006). Drosophila dorsal paired medial neurons provide a general mechanism for memory consolidation. Curr Biol *16*, 1524-1530.

Keene, A.C., Stratmann, M., Keller, A., Perrat, P.N., Vosshall, L.B., and Waddell, S. (2004). Diverse odor-conditioned memories require uniquely timed dorsal paired medial neuron output. Neuron *44*, 521-533.

Keene, A.C., and Waddell, S. (2007). Drosophila olfactory memory: single genes to complex neural circuits. Nat Rev Neurosci *8*, 341-354.

Keller, A., Sweeney, S.T., Zars, T., O'Kane, C.J., and Heisenberg, M. (2002). Targeted expression of tetanus neurotoxin interferes with behavioral responses to sensory input in Drosophila. J Neurobiol *50*, 221-233.

Kidokoro, Y., Kuromi, H., Delgado, R., Maureira, C., Oliva, C., and Labarca, P. (2004). Synaptic vesicle pools and plasticity of synaptic transmission at the Drosophila synapse. Brain Res Brain Res Rev *47*, 18-32.

Kim, Y.C., Lee, H.G., and Han, K.A. (2007). D1 dopamine receptor dDA1 is required in the mushroom body neurons for aversive and appetitive learning in Drosophila. J Neurosci *27*, 7640-7647.

Kim, Y.C., Lee, H.G., Seong, C.S., and Han, K.A. (2003). Expression of a D1 dopamine receptor dDA1/DmDOP1 in the central nervous system of Drosophila melanogaster. Gene Expr Patterns *3*, 237-245.

Kitamoto, T. (2001). Conditional modification of behavior in Drosophila by targeted expression of a temperature-sensitive shibire allele in defined neurons. J Neurobiol *47*, 81-92.

Kitamoto, T. (2002). Conditional disruption of synaptic transmission induces male-male courtship behavior in Drosophila. Proc Natl Acad Sci U S A *99*, 13232-13237.

Kittel, R.J., Wichmann, C., Rasse, T.M., Fouquet, W., Schmidt, M., Schmid, A., Wagh, D.A., Pawlu, C., Kellner, R.R., Willig, K.I., et al. (2006). Bruchpilot promotes active zone assembly, Ca2+ channel clustering, and vesicle release. Science *312*, 1051-1054.

Klagges, B.R., Heimbeck, G., Godenschwege, T.A., Hofbauer, A., Pflugfelder, G.O., Reifegerste, R., Reisch, D., Schaupp, M., Buchner, S., and Buchner, E. (1996). Invertebrate synapsins: a single gene codes for several isoforms in Drosophila. J Neurosci *16*, 3154-3165.

Knapek, S., Gerber, B., and Tanimoto H. (2010). Synapsin is selectively required for anesthesia-sensitive memory. Learn. Mem. *17*, 76-79.

Knapek, S., Sigrist, S., and Tanimoto, H. (2011). Bruchpilot, a synaptic active zone protein for anesthesia-resistant memory. J Neurosci *31*, 3453–3458.

Konopka, R.J., and Benzer, S. (1971). Clock mutants of Drosophila melanogaster. Proc Natl Acad Sci U S A *68*, 2112-2116.

Krashes, M.J., Keene, A.C., Leung, B., Armstrong, J.D., and Waddell, S. (2007). Sequential use of mushroom body neuron subsets during drosophila odor memory processing. Neuron *53*, 103-115.

Krashes, M.J., and Waddell, S. (2008). Rapid consolidation to a radish and protein synthesis-dependent long-term memory after single-session appetitive olfactory conditioning in Drosophila. J Neurosci *28*, 3103-3113.

Kuromi, H., and Kidokoro, Y. (2000). Tetanic stimulation recruits vesicles from reserve pool via a cAMP-mediated process in Drosophila synapses. Neuron *27*, 133-143.

Kuromi, H., and Kidokoro, Y. (2002). Selective replenishment of two vesicle pools depends on the source of Ca2+ at the Drosophila synapse. Neuron *35*, 333-343.

Lee, H.G., Seong, C.S., Kim, Y.C., Davis, R.L., and Han, K.A. (2003). Octopamine receptor OAMB is required for ovulation in Drosophila melanogaster. Dev Biol *264*, 179-190.

Lee, T., Lee, A., and Luo, L. (1999). Development of the Drosophila mushroom bodies: sequential generation of three distinct types of neurons from a neuroblast. Development *126*, 4065-4076.

Levin, L.R., Han, P.L., Hwang, P.M., Feinstein, P.G., Davis, R.L., and Reed, R.R. (1992). The Drosophila learning and memory gene rutabaga encodes a Ca2+/Calmodulin-responsive adenylyl cyclase. Cell *68*, 479-489.

Li, L., Chin, L.S., Shupliakov, O., Brodin, L., Sihra, T.S., Hvalby, O., Jensen, V., Zheng, D., McNamara, J.O., Greengard, P., and et al. (1995). Impairment of synaptic vesicle clustering and of synaptic transmission, and increased seizure propensity, in synapsin I-deficient mice. Proc Natl Acad Sci U S A *92*, 9235-9239.

Li, W., Tully, T., and Kalderon, D. (1996). Effects of a conditional Drosophila PKA mutant on olfactory learning and memory. Learn Mem *2*, 320-333.

Lin, H.H., Lai, J.S., Chin, A.L., Chen, Y.C., and Chiang, A.S. (2007). A map of olfactory representation in the Drosophila mushroom body. Cell *128*, 1205-1217.

Lin, X., Ruan, X., Anderson, M.G., McDowell, J.A., Kroeger, P.E., Fesik, S.W., and Shen, Y. (2005). siRNA-mediated off-target gene silencing triggered by a 7 nt complementation. Nucleic Acids Res *33*, 4527-4535.

Ma, Y., Creanga, A., Lum, L., and Beachy, P.A. (2006). Prevalence of off-target effects in Drosophila RNA interference screens. Nature *443*, 359-363.

Mao, Z., and Davis, R.L. (2009). Eight different types of dopaminergic neurons innervate the Drosophila mushroom body neuropil: anatomical and physiological heterogeneity. Front Neural Circuits *3*, 5.

Mao, Z., Roman, G., Zong, L., and Davis, R.L. (2004). Pharmacogenetic rescue in time and space of the rutabaga memory impairment by using Gene-Switch. Proc Natl Acad Sci U S A *101*, 198-203.

Maqueira, B., Chatwin, H., and Evans, P.D. (2005). Identification and characterization of a novel family of Drosophila beta-adrenergic-like octopamine G-protein coupled receptors. J Neurochem *94*, 547-560.

Margulies, C., Tully, T., and Dubnau, J. (2005). Deconstructing memory in Drosophila. Curr Biol *15*, R700-713.

Marin, E.C., Jefferis, G.S., Komiyama, T., Zhu, H., and Luo, L. (2002). Representation of the glomerular olfactory map in the Drosophila brain. Cell *109*, 243-255.

Masek, P., and Heisenberg, M. (2008). Distinct memories of odor intensity and quality in Drosophila. Proc Natl Acad Sci U S A *105*, 15985-15990.

McGaugh, J.L. (2000). Memory--a century of consolidation. Science *287*, 248-251.

McGuire, S.E., Deshazer, M., and Davis, R.L. (2005). Thirty years of olfactory learning and memory research in Drosophila melanogaster. Prog Neurobiol *76*, 328-347.

McGuire, S.E., Le, P.T., and Davis, R.L. (2001). The role of Drosophila mushroom body signaling in olfactory memory. Science 293, 1330-1333.

McGuire, S.E., Le, P.T., Osborn, A.J., Matsumoto, K., and Davis, R.L. (2003). Spatiotemporal rescue of memory dysfunction in Drosophila. Science 302, 1765-1768.

Menegon, A., Bonanomi, D., Albertinazzi, C., Lotti, F., Ferrari, G., Kao, H.T., Benfenati, F., Baldelli, P., and Valtorta, F. (2006). Protein kinase A-mediated synapsin I phosphorylation is a central modulator of Ca2+-dependent synaptic activity. J Neurosci 26, 11670-11681.

Menzel, R. (2001). Searching for the memory trace in a mini-brain, the honeybee. Learn Mem 8, 53-62.

Mery, F., and Kawecki, T.J. (2005). A cost of long-term memory in Drosophila. Science 308, 1148.

Michels, B., Diegelmann, S., Tanimoto, H., Schwenkert, I., Buchner, E., and Gerber, B. (2005). A role for Synapsin in associative learning: the Drosophila larva as a study case. Learn Mem 12, 224-231.

Mizunami, M., Unoki, S., Mori, Y., Hirashima, D., Hatano, A., and Matsumoto, Y. (2009). Roles of octopaminergic and dopaminergic neurons in appetitive and aversive memory recall in an insect. BMC Biol 7, 46.

Moffat, J., Reiling, J.H., and Sabatini, D.M. (2007). Off-target effects associated with long dsRNAs in Drosophila RNAi screens. Trends Pharmacol Sci 28, 149-151.

Monastirioti, M., Linn, C.E., Jr., and White, K. (1996). Characterization of Drosophila tyramine beta-hydroxylase gene and isolation of mutant flies lacking octopamine. J Neurosci 16, 3900-3911.

Moore, M.S., DeZazzo, J., Luk, A.Y., Tully, T., Singh, C.M., and Heberlein, U. (1998). Ethanol intoxication in Drosophila: Genetic and pharmacological evidence for regulation by the cAMP signaling pathway. Cell 93, 997-1007.

Morgan, T.H. (1910). Sex Limited Inheritance in Drosophila. Science 32, 120-122.

Ng, M., Roorda, R.D., Lima, S.Q., Zemelman, B.V., Morcillo, P., and Miesenbock, G. (2002). Transmission of olfactory information between three populations of neurons in the antennal lobe of the fly. Neuron 36, 463-474.

Nighorn, A., Healy, M.J., and Davis, R.L. (1991). The cyclic AMP phosphodiesterase encoded by the Drosophila dunce gene is concentrated in the mushroom body neuropil. Neuron 6, 455-467.

Novoseltsev, V.N., Arking, R., Carey, J.R., Novoseltseva, J.A., and Yashin, A.I. (2005). Individual fecundity and senescence in Drosophila and medfly. J Gerontol A Biol Sci Med Sci 60, 953-962.

O'Kane, C.J., and Gehring, W.J. (1987). Detection in situ of genomic regulatory elements in Drosophila. Proc Natl Acad Sci U S A *84*, 9123-9127.

Pan, Y., Zhou, Y., Guo, C., Gong, H., Gong, Z., and Liu, L. (2009). Differential roles of the fan-shaped body and the ellipsoid body in Drosophila visual pattern memory. Learn Mem *16*, 289-295.

Pascual, A., and Preat, T. (2001). Localization of long-term memory within the Drosophila mushroom body. Science *294*, 1115-1117.

Pavlov, I.P. (1927). Conditioned Reflexes (London: Oxford University Press).

Pieribone, V.A., Shupliakov, O., Brodin, L., Hilfiker-Rothenfluh, S., Czernik, A.J., and Greengard, P. (1995). Distinct pools of synaptic vesicles in neurotransmitter release. Nature *375*, 493-497.

Preat, T. (1998). Decreased odor avoidance after electric shock in Drosophila mutants biases learning and memory tests. J Neurosci *18*, 8534-8538.

Quinn, W.G., and Dudai, Y. (1976). Memory phases in Drosophila. Nature *262*, 576-577.

Quinn, W.G., Harris, W.A., and Benzer, S. (1974). Conditioned behavior in Drosophila melanogaster. Proc Natl Acad Sci U S A *71*, 708-712.

Quinn, W.G., Sziber, P.P., and Booker, R. (1979). The Drosophila memory mutant amnesiac. Nature *277*, 212-214.

Rein, K., Zockler, M., Mader, M.T., Grubel, C., and Heisenberg, M. (2002). The Drosophila standard brain. Curr Biol *12*, 227-231.

Reiter, L.T., Potocki, L., Chien, S., Gribskov, M., and Bier, E. (2001). A systematic analysis of human disease-associated gene sequences in Drosophila melanogaster. Genome Res *11*, 1114-1125.

Riemensperger, T., Voller, T., Stock, P., Buchner, E., and Fiala, A. (2005). Punishment prediction by dopaminergic neurons in Drosophila. Curr Biol *15*, 1953-1960.

Roman, G. (2004). The genetics of Drosophila transgenics. Bioessays *26*, 1243-1253.

Rosahl, T.W., Geppert, M., Spillane, D., Herz, J., Hammer, R.E., Malenka, R.C., and Sudhof, T.C. (1993). Short-term synaptic plasticity is altered in mice lacking synapsin I. Cell *75*, 661-670.

Rosahl, T.W., Spillane, D., Missler, M., Herz, J., Selig, D.K., Wolff, J.R., Hammer, R.E., Malenka, R.C., and Sudhof, T.C. (1995). Essential functions of synapsins I and II in synaptic vesicle regulation. Nature *375*, 488-493.

Rubin, G.M., Yandell, M.D., Wortman, J.R., Gabor Miklos, G.L., Nelson, C.R., Hariharan, I.K., Fortini, M.E., Li, P.W., Apweiler, R., Fleischmann, W., et al. (2000). Comparative genomics of the eukaryotes. Science 287, 2204-2215.

Scholz, H., Ramond, J., Singh, C.M., and Heberlein, U. (2000). Functional ethanol tolerance in Drosophila. Neuron 28, 261-271.

Schroll, C., Riemensperger, T., Bucher, D., Ehmer, J., Voller, T., Erbguth, K., Gerber, B., Hendel, T., Nagel, G., Buchner, E., and Fiala, A. (2006). Light-induced activation of distinct modulatory neurons triggers appetitive or aversive learning in Drosophila larvae. Curr Biol 16, 1741-1747.

Schwaerzel, M., Heisenberg, M., and Zars, T. (2002). Extinction antagonizes olfactory memory at the subcellular level. Neuron 35, 951-960.

Schwaerzel, M., Jaeckel, A., and Mueller, U. (2007). Signaling at A-kinase anchoring proteins organizes anesthesia-sensitive memory in Drosophila. J Neurosci 27, 1229-1233.

Schwaerzel, M., Monastirioti, M., Scholz, H., Friggi-Grelin, F., Birman, S., and Heisenberg, M. (2003). Dopamine and octopamine differentiate between aversive and appetitive olfactory memories in Drosophila. J Neurosci 23, 10495-10502.

Selcho, M., Pauls, D., Han, K.A., Stocker, R.F., and Thum, A.S. (2009). The role of dopamine in Drosophila larval classical olfactory conditioning. PLoS One 4, e5897.

Sherff, C.M., and Carew, T.J. (2004). Parallel somatic and synaptic processing in the induction of intermediate-term and long-term synaptic facilitation in Aplysia. Proc Natl Acad Sci U S A 101, 7463-7468.

Silva, A.J., Rosahl, T.W., Chapman, P.F., Marowitz, Z., Friedman, E., Frankland, P.W., Cestari, V., Cioffi, D., Sudhof, T.C., and Bourtchuladze, R. (1996). Impaired learning in mice with abnormal short-lived plasticity. Curr Biol 6, 1509-1518.

Sinakevitch, I., and Strausfeld, N.J. (2006). Comparison of octopamine-like immunoreactivity in the brains of the fruit fly and blow fly. J Comp Neurol 494, 460-475.

Skinner, B.F. (1938). The Behavior of Organisms (New York: D. Appleton-Century Company).

Skoulakis, E.M., and Davis, R.L. (1996). Olfactory learning deficits in mutants for leonardo, a Drosophila gene encoding a 14-3-3 protein. Neuron 17, 931-944.

Skoulakis, E.M., Kalderon, D., and Davis, R.L. (1993). Preferential expression in mushroom bodies of the catalytic subunit of protein kinase A and its role in learning and memory. Neuron 11, 197-208.

Stocker, R.F., Lienhard, M.C., Borst, A., and Fischbach, K.F. (1990). Neuronal architecture of the antennal lobe in Drosophila melanogaster. Cell Tissue Res 262, 9-34.

Strausfeld, N.J., Hansen, L., Li, Y., Gomez, R.S., and Ito, K. (1998). Evolution, discovery, and interpretations of arthropod mushroom bodies. Learn Mem 5, 11-37.

Strausfeld, N.J., Sinakevitch, I., and Vilinsky, I. (2003). The mushroom bodies of Drosophila melanogaster: an immunocytological and golgi study of Kenyon cell organization in the calyces and lobes. Microsc Res Tech 62, 151-169.

Strauss, R., Hanesch, U., Kinkelin, M., Wolf, R., and Heisenberg, M. (1992). No-bridge of Drosophila melanogaster: portrait of a structural brain mutant of the central complex. J Neurogenet 8, 125-155.

Sun, J., Bronk, P., Liu, X., Han, W., and Sudhof, T.C. (2006). Synapsins regulate use-dependent synaptic plasticity in the calyx of Held by a Ca2+/calmodulin-dependent pathway. Proc Natl Acad Sci U S A 103, 2880-2885.

Sweeney, S.T., Broadie, K., Keane, J., Niemann, H., and O'Kane, C.J. (1995). Targeted expression of tetanus toxin light chain in Drosophila specifically eliminates synaptic transmission and causes behavioral defects. Neuron 14, 341-351.

Tamura, T., Chiang, A.S., Ito, N., Liu, H.P., Horiuchi, J., Tully, T., and Saitoe, M. (2003). Aging specifically impairs amnesiac-dependent memory in Drosophila. Neuron 40, 1003-1011.

Tanaka, N.K., Awasaki, T., Shimada, T., and Ito, K. (2004). Integration of chemosensory pathways in the Drosophila second-order olfactory centers. Curr Biol 14, 449-457.

Tanaka, N.K., Tanimoto, H., and Ito, K. (2008). Neuronal assemblies of the Drosophila mushroom body. J Comp Neurol 508, 711-755.

Tang, S., and Guo, A. (2001). Choice behavior of Drosophila facing contradictory visual cues. Science 294, 1543-1547.

Technau, G., and Heisenberg, M. (1982). Neural reorganization during metamorphosis of the corpora pedunculata in Drosophila melanogaster. Nature 295, 405-407.

Tempel, B.L., Bonini, N., Dawson, D.R., and Quinn, W.G. (1983). Reward learning in normal and mutant Drosophila. Proc Natl Acad Sci U S A 80, 1482-1486.

Thum, A.S. (2006). Sugar reward learning in Drosophila. Neuronal circuits in Drosophila associative olfactory learning. In Lehrstuhl für Genetik und Neurobiologie (Würzburg, Bayerische Julius-Maximilians-Universität).

Thum, A.S., Jenett, A., Ito, K., Heisenberg, M., and Tanimoto, H. (2007). Multiple memory traces for olfactory reward learning in Drosophila. J Neurosci 27, 11132-11138.

Tully, T., Preat, T., Boynton, S.C., and Del Vecchio, M. (1994). Genetic dissection of consolidated memory in Drosophila. Cell 79, 35-47.

Tully, T., and Quinn, W.G. (1985). Classical conditioning and retention in normal and mutant Drosophila melanogaster. J Comp Physiol A *157*, 263-277.

Unoki, S., Matsumoto, Y., and Mizunami, M. (2005). Participation of octopaminergic reward system and dopaminergic punishment system in insect olfactory learning revealed by pharmacological study. Eur J Neurosci *22*, 1409-1416.

Vergoz, V., Roussel, E., Sandoz, J.C., and Giurfa, M. (2007). Aversive learning in honeybees revealed by the olfactory conditioning of the sting extension reflex. PLoS One *2*, e288.

Villella, A., Gailey, D.A., Berwald, B., Ohshima, S., Barnes, P.T., and Hall, J.C. (1997). Extended reproductive roles of the fruitless gene in Drosophila melanogaster revealed by behavioral analysis of new fru mutants. Genetics *147*, 1107-1130.

Waddell, S., Armstrong, J.D., Kitamoto, T., Kaiser, K., and Quinn, W.G. (2000). The amnesiac gene product is expressed in two neurons in the Drosophila brain that are critical for memory. Cell *103*, 805-813.

Wagh, D.A., Rasse, T.M., Asan, E., Hofbauer, A., Schwenkert, I., Durrbeck, H., Buchner, S., Dabauvalle, M.C., Schmidt, M., Qin, G., *et al.* (2006). Bruchpilot, a protein with homology to ELKS/CAST, is required for structural integrity and function of synaptic active zones in Drosophila. Neuron *49*, 833-844.

Wang, Y., Guo, H.F., Pologruto, T.A., Hannan, F., Hakker, I., Svoboda, K., and Zhong, Y. (2004). Stereotyped odor-evoked activity in the mushroom body of Drosophila revealed by green fluorescent protein-based Ca2+ imaging. J Neurosci *24*, 6507-6514.

White, K., Tahaoglu, E., and Steller, H. (1996). Cell killing by the Drosophila gene reaper. Science *271*, 805-807.

Williams, R.W., and Herrup, K. (1988). The control of neuron number. Annu Rev Neurosci *11*, 423-453.

Wilson, R.I., and Laurent, G. (2005). Role of GABAergic inhibition in shaping odor-evoked spatiotemporal patterns in the Drosophila antennal lobe. J Neurosci *25*, 9069-9079.

Wong, A.M., Wang, J.W., and Axel, R. (2002). Spatial representation of the glomerular map in the Drosophila protocerebrum. Cell *109*, 229-241.

Xia, S., Miyashita, T., Fu, T.F., Lin, W.Y., Wu, C.L., Pyzocha, L., Lin, I.R., Saitoe, M., Tully, T., and Chiang, A.S. (2005). NMDA receptors mediate olfactory learning and memory in Drosophila. Curr Biol *15*, 603-615.

Xia, S., and Tully, T. (2007). Segregation of odor identity and intensity during odor discrimination in Drosophila mushroom body. PLoS Biol *5*, e264.

Yang, M.Y., Armstrong, J.D., Vilinsky, I., Strausfeld, N.J., and Kaiser, K. (1995). Subdivision of the Drosophila mushroom bodies by enhancer-trap expression patterns. Neuron *15*, 45-54.

Yarali, A., Ehser, S., Hapil, F.Z., Huang, J., and Gerber, B. (2009). Odour intensity learning in fruit flies. Proc Biol Sci *276*, 3413-3420.

Yu, D., Akalal, D.B., and Davis, R.L. (2006). Drosophila alpha/beta mushroom body neurons form a branch-specific, long-term cellular memory trace after spaced olfactory conditioning. Neuron *52*, 845-855.

Yu, D., Keene, A.C., Srivatsan, A., Waddell, S., and Davis, R.L. (2005). Drosophila DPM neurons form a delayed and branch-specific memory trace after olfactory classical conditioning. Cell *123*, 945-957.

Yu, D., Ponomarev, A., and Davis, R.L. (2004). Altered representation of the spatial code for odors after olfactory classical conditioning; memory trace formation by synaptic recruitment. Neuron *42*, 437-449.

Zars, T. (2000). Behavioral functions of the insect mushroom bodies. Curr Opin Neurobiol *10*, 790-795.

Zars, T., Fischer, M., Schulz, R., and Heisenberg, M. (2000). Localization of a short-term memory in Drosophila. Science *288*, 672-675.

Zhang, K., Guo, J.Z., Peng, Y., Xi, W., and Guo, A. (2007). Dopamine-mushroom body circuit regulates saliency-based decision-making in Drosophila. Science *316*, 1901-1904.

6   List of abbreviations

AC	adenylyl cyclase
aimpr	anterior inferior medial protocerebrum
AL	antennal lobe
amn	*amnesiac*
ARM	anesthesia-resistant memory
ASM	anesthesia-sensitive memory
AZ	active zone
brp	*bruchpilot*
cAMP	cyclic adenosine monophosphate
CR	conditioned response
CS	conditioned stimulus
dnc	*dunce*
DPM	dorsal paired medial
dsRNA	double-stranded RNA
GAL4	yeast transcription factor
GAL80	silencer of GAL4
GFP	green fluorescent protein
h	hour
iACT	inner antennocerebral tract
LTM	long-term memory
mACT	medial antennocerebral tract
MB	mushroom body
min	minute
mRNA	messenger RNA
MTM	middle-term memory
n	number of experiments
ns	not significant

List of abbreviation

OIM	odor intensity memory
OR	olfactory receptor
ORN	olfactory receptor neuron
PKA	protein kinase A
PN	projection neuron
PSD	postsynaptic density
rad	*radish*
RISC	RNA-induced silencing complex
RNAi	RNA interference
RP	reserve pool
RRP	readily releasable pool
rut	*rutabaga*
shi^{ts}	temperature-sensitive, dominant-negative allele of *shibire*
siRNA	small interfering RNA
STM	short-term memory
syn	*synapsin*
TβH	tyramine-β-hydroxylase
UAS	upstream activating sequence
UR	unconditioned response
US	unconditioned stimulus
V	volt
VUM	ventral unpaired medial

7 Summary

Memory is dynamic: shortly after acquisition it is susceptible to amnesic treatments, gets gradually consolidated, and becomes resistant to retrograde amnesia (McGaugh, 2000). Associative olfactory memory of the fruit fly *Drosophila melanogaster* also shows these features. After a single associative training where an odor is paired with electric shock (Quinn et al., 1974; Tully and Quinn, 1985), flies form an aversive odor memory that lasts for several hours, consisting of qualitatively different components. These components can be dissociated by mutations, their underlying neuronal circuitry and susceptibility to amnesic treatments (Dubnau and Tully, 1998; Isabel et al., 2004; Keene and Waddell, 2007; Masek and Heisenberg, 2008; Xia and Tully, 2007). A component that is susceptible to an amnesic treatment, i.e. anesthesia-sensitive memory (ASM), dominates early memory, but decays rapidly (Margulies et al., 2005; Quinn and Dudai, 1976). A consolidated anesthesia-resistant memory component (ARM) is built gradually within the following hours and lasts significantly longer (Margulies et al., 2005; Quinn and Dudai, 1976). I showed here that the establishment of ARM requires less intensity of shock reinforcement than ASM.

ARM and ASM rely on different molecular and/or neuronal processes: ARM is selectively impaired in the *radish* mutant, whereas for example the *amnesiac* and *rutabaga* genes are specifically required for ASM (Dudai et al., 1988; Folkers et al., 1993; Isabel et al., 2004; Quinn and Dudai, 1976; Schwaerzel et al., 2007; Tully et al., 1994). The latter comprise the cAMP signaling pathway in the fly, with the PKA being its supposed major target (Levin et al., 1992). Here I showed that a *synapsin* null-mutant encoding the evolutionary conserved phosphoprotein Synapsin is selectively impaired in the labile ASM. Further experiments suggested Synapsin as a potential downstream effector of the cAMP/PKA cascade. Similar to my results, Synapsin plays a role for different learning tasks in vertebrates (Gitler et al., 2004; Silva et al., 1996). Also in *Aplysia*, PKA-dependent phosphorylation of Synapsin has been proposed to be involved in regulation of neurotransmitter release and short-term plasticity (Angers et al., 2002; Fiumara et al., 2004).

Synapsin is associated with a reserve pool of vesicles at the presynapse and is required to maintain vesicle release specifically under sustained high frequency nerve stimulation (Akbergenova and Bykhovskaia, 2007; Li et al., 1995; Pieribone et al., 1995; Sun et al., 2006). In contrast, the requirement of Bruchpilot, which is homologous to the mammalian active zone proteins ELKS/CAST (Wagh et al., 2006), is most pronounced in immediate vesicle release (Kittel et al., 2006). Under repeated stimulation of a *bruchpilot* mutant motor neuron, immediate vesicle release is severely impaired whereas the following steady-state release is still possible (Kittel et al., 2006). In line with that, knockdown of the Bruchpilot protein causes impairment in clustering of Ca^{2+} channels to the active zones and a lack of electron-dense projections at presynaptic terminals (T-bars). Thus, less synaptic vesicles of the readily-releasable pool are accumulated to the release sites and their release probability is severely impaired (Kittel et al., 2006; Wagh et al., 2006). First, I showed that Bruchpilot is required for aversive olfactory memory and localized the requirement of Bruchpilot to the Kenyon cells of the mushroom body, the second-order olfactory interneurons in *Drosophila*. Furthermore, I demonstrated that Bruchpilot selectively functions for the consolidated anesthesia-resistant memory. Since Synapsin is specifically required for the labile anesthesia sensitive memory, different synaptic proteins can dissociate consolidated and labile components of olfactory memory and two different modes of neurotransmission (high- vs. low frequency dependent) might differentiate ASM and ARM.

8 Zusammenfassung

Gedächtnis ist ein dynamischer Prozess. In der Zeit kurz nach seiner Bildung ist es instabil und anfällig gegen amnestische Störungen, dann wird es schrittweise konsolidiert und schließlich resistent gegenüber retrogradem Gedächtnisverlust (McGaugh, 2000). Auch das assoziative olfaktorische Gedächtnis der Fruchtfliege *Drosophila melanogaster* zeigt diese Merkmale. Nach einem einzelnen assoziativen Training, in welchem ein Duft mit elektrischen Stromstößen gepaart wird, bilden die Fliegen ein aversives Duftgedächtnis, welches über mehrere Stunden anhält und aus qualitativ unterschiedlichen Komponenten besteht (Quinn et al., 1974; Tully and Quinn, 1985). Diese Komponenten können zum Beispiel durch Mutationen, die zugrunde liegenden neuronalen Verknüpfungen oder durch ihre Anfälligkeit für amnestische Behandlungen unterschieden werden (Dubnau and Tully, 1998; Isabel et al., 2004; Keene and Waddell, 2007; Masek and Heisenberg, 2008; Xia and Tully, 2007). Eine gegen amnestische Behandlungen, wie beispielsweise Kälte-induzierte Betäubung, anfällige Komponente beherrscht das frühe Gedächtnis, zerfällt jedoch schnell (Margulies et al., 2005; Quinn and Dudai, 1976). Diese wird deshalb Anästhesie-sensitives Gedächtnis genannt (anesthesia-sensitive memory [ASM]). Im Gegensatz dazu baut sich eine konsolidierte Komponente erst langsam in den folgenden Stunden nach dem Training auf, hält stattdessen jedoch länger an (Margulies et al., 2005; Quinn and Dudai, 1976). Diese Komponente ist resistent gegenüber Kälte-induzierter Anästhesie und wird deshalb als ARM (anesthesia-resistant memory) bezeichnet. In der vorliegenden Arbeit konnte ich zeigen, dass das konsolidierte ARM bereits mit deutlich weniger starken Elektroschocks im Training gebildet wird als das instabile ASM.

ARM und ASM unterliegen unterschiedliche molekulare und/oder neuronale Prozesse. Während in einer Mutante für das *radish* Gen selektiv ARM beeinträchtigt ist, werden andere Gene wie zum Beispiel *amnesiac* oder *rutabaga* ausschließlich für ASM benötigt (Dudai et al., 1988; Folkers et al., 1993; Isabel et al., 2004; Quinn and Dudai, 1976; Schwaerzel et al., 2007; Tully et al., 1994). Die beiden letzteren sind Teil des cAMP Signalweges, welcher vermutlich hauptsächlich die cAMP abhängige Protein-Kinase A (PKA) aktiviert (Levin et al., 1992). Hier zeige ich, dass eine Null-Mutante für das evolutionär konservierte Phosphoprotein Synapsin einen selektiven Defekt in ASM hat. Weitere Experimente lassen vermuten, dass Synapsin als Effektor stromabwärts der cAMP/PKA Kaskade wirkt. Ähnlich wie bei *Drosophila* spielt Synaspin auch in Vertebraten eine Rolle in unterschiedlichen Lernparadigmen (Gitler et al., 2004; Silva et al., 1996). Auch in der Meeresschnecke *Aplysia* wurde eine PKA abhängige Phosphorylierung von Synapsin als Mechanismus für die Regulierung von Neurotransmitterausschüttung und Kurzzeitplastizität vorgeschlagen (Angers et al., 2002; Fiumara et al., 2004).

Synapsin wird für die Bildung eines Reserve-Pools von Vesikeln an der Präsynapse und für die Aufrechterhaltung der Vesikelausschüttung speziell bei anhaltender, hochfrequenter Stimulation von Nervenzellen benötigt (Akbergenova and Bykhovskaia, 2007; Li et al., 1995; Pieribone et al., 1995; Sun et al., 2006). Im Gegensatz dazu wird Bruchpilot, ein Protein der aktiven Zone und homolog zu den ELKS/CAST Proteinen bei Säugern (Wagh et al., 2006), haupsächlich für sofortige Vesikelausschüttung gebraucht (Kittel et al., 2006). Bei wiederholter Stimulation an Motorneuronen einer *bruchpilot* Mutante ist die akute Vesikelausschüttung stark vermindert, während die darauf folgende andauernde Ausschüttung noch immer möglich ist (Kittel et al., 2006). Dazu passend beeinträchtigt eine künstliche Verminderung des Bruchpilot-Proteins die Ansammlung von Ca^{2+} Kanälen an den aktiven Zonen, sowie die Bildung von elektronendichten Strukturen (T-bars) an den präsynaptischen Endigungen. Deshalb akkumulieren weniger Vesikel des "readily-releasable" Pools an den Ausschüttungsstellen und die Ausschüttungswahrscheinlichkeit ist stark vermindert (Kittel et al., 2006; Wagh et al., 2006). In dieser Arbeit zeige ich zum ersten Mal, dass Bruchpilot für aversives olfaktorisches Gedächtnis benötigt wird. Der Ort an dem Bruchpilot hierfür gebraucht wird sind die Kenyon-Zellen des Pilzkörpers, die olfaktorischen Interneuronen zweiter Ordnung in *Drosophila*. Desweiteren zeige ich, dass die Funktion von Bruchpilot selektiv für das konsolidierte ARM ist. Da Synapsin spezifisch für das labile ASM benötigt wird, können diese beiden olfaktorischen Gedächtniskomponenten durch verschiedene synaptische Proteine getrennt werden, und zwei unterschiedliche Arten der Neurotransmitterausschüttung (abhängig von hoch- oder niedrigfrequenter Stimulation) könnten ASM und ARM auseinander halten.

I want morebooks!

Buy your books fast and straightforward online - at one of world's fastest growing online book stores! Environmentally sound due to Print-on-Demand technologies.

Buy your books online at
www.morebooks.shop

Kaufen Sie Ihre Bücher schnell und unkompliziert online – auf einer der am schnellsten wachsenden Buchhandelsplattformen weltweit! Dank Print-On-Demand umwelt- und ressourcenschonend produziert.

Bücher schneller online kaufen
www.morebooks.shop

KS OmniScriptum Publishing
Brivibas gatve 197
LV-1039 Riga, Latvia
Telefax +371 686 204 55

info@omniscriptum.com
www.omniscriptum.com

Printed by Books on Demand GmbH, Norderstedt / Germany